COMPLEXITY

COMPLEXITY

The Evolution *of* Earth's Biodiversity
and the Future *of* Humanity

WILLIAM C. BURGER

PB Prometheus Books

59 John Glenn Drive
Amherst, New York 14228

Published 2016 by Prometheus Books

Cover images © Brand X Pictures
Cover design by Nicole Sommer-Lecht

Inquiries should be addressed to
Prometheus Books
59 John Glenn Drive
Amherst, New York 14228
VOICE: 716–691–0133
FAX: 716–691–0137
WWW.PROMETHEUSBOOKS.COM

20 19 18 17 16   5 4 3 2 1

Library of Congress Cataloging-in-Publication Data

Names: Burger, William C.
Title: Complexity : the evolution of Earth's biodiversity and the future of
    humanity / by William C. Burger.
Description: Amherst, New York : Prometheus Books, 2016. |
    Includes bibliographical references and index.
Identifiers: LCCN 2016007382 (print) | LCCN 2016017156 (ebook) |
    ISBN 9781633881938 (hardcover) | ISBN 9781633881945 (ebook)
Subjects: LCSH: Biodiversity. | Biotic communities. | Macroecology. |
    Environmental health.
Classification: LCC QH541.15.B56 B88 2016 (print) | LCC QH541.15.B56 (ebook) |
    DDC 333.95—dc23
LC record available at https://lccn.loc.gov/2016007382

Printed in the United States of America

CONTENTS

INTRODUCTION

Planet Earth abounds in rich diversity. Armies of insects, an abundance of birds, and flowers blooming around the world are only the more obvious manifestations of this grand variety.[1] More significantly, plant and animal life seem to have become both more diverse and more complex over geological time. Today, our own species, already replete with many cultures and thousands of languages, is adding still more to the world's complexity with ever-expanding technologies. In fact, human success has been fostered by the natural diversity that surrounds us. In the final chapters, we will discuss how our own species has contributed to global complexity, but let's begin with natural diversity itself: how it is measured, how it is distributed, and how it got that way.

Trying to understand biodiversity and the processes that allow so many organisms to co-exist has been a challenge for the biological sciences over the last few hundred years. Sciences that call themselves ecology, biogeography, systematics, and phylogenetics are all attempts to comprehend the living world's rich variety. The subject of biodiversity—a convenient abbreviation for biological diversity—spans all these different disciplines. In addition, paleontology shows us how biodiversity has grown and expanded over geological time, while conservation biology examines ways of preserving our natural heritage. Surely human success would have been impossible without a foundation of biological abundance.

Considered in its widest sense, biodiversity includes genetic diversity and human cultural variety, as well as the interactions shared by all living things. The idea of diversity—the numbers of things—is a little different from our notions of complexity, in that complexity depends on the *intricacy and number of parts* comprising a particular system. Thus, beetles are more complex than jellyfish, while the rain forest is more complex than the tundra. And, as if all this weren't amazing enough, we humans are adding ever greater complexity to the planet ourselves. But starting with the nonhuman world, what are we talking about when we discuss biodiversity?

COUNTING SPECIES

Whether animal or vegetable, species are the common currency in discussing biodiversity. Species numbers are the first statistic we look at when considering a habitat's diversity. Each species has a limited range, constrained by its own specific physiological requirements. While some species may range across a continent, others are confined to very specific localities. All together, they make up the biota that characterizes a particular site or region. But how do we recognize and characterize the **biological species**? A useful working definition must be able to identify populations in the field and also permit comparisons around the world. In general, members of the same wild species share genetic information by interbreeding with others of their own kind, *and not with other species*! Because they don't exchange genes with other species, individual species can be distinguished from closely related species. We'll discuss this in more detail in chapter 3, but the *species* does seem to be the most fundamental and realistic element in our hierarchy of taxonomic categories. Folk taxonomies often agree with their scientific counterparts, supporting the reality of our species concepts.[2] Studying birds in the Arfak mountains of New Guinea, Ernst Mayr found that the local people recognized 136 kinds; after careful study, he came to recognize 137 species. Most species do seem to be "real" and not just a matter of opinion. More importantly, the world really does have a lot of different species.[3]

Species are the bottom rung in the hierarchy of living things. Higher categories are more inclusive, representing a larger sample of the living world. We place a group of species, which are more similar to each other than to other species, into a single genus (plural: genera), similar genera into families, similar families into orders, and similar orders into classes.

Finally, classes are placed under a larger category: divisions (for plants) or phyla (for animals). Plants are members of the kingdom **Plantae,** and animals are members of the kingdom **Animalia. Fungi,** bacteria, and various microorganisms are arranged in additional kingdoms. This is our taxonomic hierarchy, a way of cataloging and making sense of the natural world—a methodology that has proven immensely useful over these last two hundred years. Wolves, for example, are considered a single species and placed in the genus *Canis.* That genus, which includes jackals and foxes, is placed in the family Canidae, the order Mammalia, and is part of the phylum Vertebrata (animals with backbones). Each of these higher categories is a larger and more inclusive grouping. Though initiated before the notion of evolutionary change had been developed, these ranks were easily incorporated into a historical framework covering hundreds of millions of years. In an evolutionary sense, higher, more inclusive ranks (such as orders and divisions) are *older* groupings, comprising descendants from a more ancient time. The largest divisions under kingdom Animalia are the phyla, and they differ greatly from each other. Fish (Vertebrates) are very different from jellyfish (Coelenterates), and neither is closely similar to clams (Mollusks). These major phyla represent categories that separated from each other more than five hundred million years ago. Species, in contrast, are seen to be the most recent twigs on a large evolutionary tree. And it is species that we will focus on most often in this book.

Unfortunately, there's a problem in determining biodiversity by counting species: which animals or plants should be given priority in our enumerations? We have neither the time, the money, nor the ability to find and count them all. At the present time, around 1.8 million species have actually been described and published, far too many to handle in any reasonable survey.[4] Efficiency requires that we follow the principle of least effort. Actually, we have really good numbers for only a few major lineages. Forget about insects; they're too numerous, too small, too poorly sampled, and too difficult to identify. Except for butterflies, trying to compare the numbers of insect species around the world is not currently feasible. We have no choice but to rely on the numbers of better-known creatures; for terrestrial surveys these are birds, mammals, reptiles, amphibians, and the higher (vascular) plants.

Birds are especially popular as indicators of diverse locales; they are an obvious choice for two reasons. First, birds have been intensely studied; of all animals, these are the ones we know the best. Being colorful, often noisy, and active mostly by day, they've gotten a lot of attention. Because many mammals come out only in the dark of night, they've avoided the scrutiny to which birds have been subjected. Plants, despite being conspicuous and stuck in the ground, present other challenges. Identifying higher plants successfully usually requires finding flowers or fruits and seeds—easier said than done. For some tropical trees, many years may pass before they flower and fruit. Desert floras can be similarly opaque: plants flower only after sufficient rainfall, being both leafless and flowerless over much of the year. With less than generous rainfall many desert species will not flower at all. Come back during the rainy season next year and try again. For these reasons, plant identification and enumeration require many years of continuing effort. Similarly, getting good lists of birds, mammals, reptiles, or amphibians can take years of extensive mist-netting, trapping, and observation. All told, finalizing species lists for local plants and animals requires years of effort. These efforts are continuing, with new discoveries being made on a regular basis. Recent work with Australian geckos and Madagascar's frogs have made clear that there are still many "unrecognized and undescribed" species to be added to our inventories.[5]

THE POWER BEHIND BIOLOGICAL DIVERSITY

Generally speaking, plant diversity is fundamental to biodiversity—and for two reasons. By transforming the physical energy of sunlight into chemical energy, plants support the local food chain. Living things cannot survive and reproduce without the timely acquisition of energy. Photosynthesis by plants, algae, and Cyanobacteria (once called blue-green algae) are the source of this energy. Within the sea, microscopic phytoplankton capture most of the Sun's energy, while on land, green plants are the foundation of the food chain.

In addition to capturing energy, larger plants provide terrestrial environments much of their structural architecture. Bird diversity, in particular, is strongly constrained by the variety of nesting sites, and such sites are dependent on the numbers of trees and shrubs. Also, with more varied plant species there will be a greater variety of fruits, seeds, and fresh green shoots, exactly what many insects, birds, and mammals are looking for. In most cases, the numbers of land animals and higher plant species will be closely congruent in any given habitat.[6]

Habitats are where biodiversity is found, and we will be referring to these as different *biomes* throughout the book. A **biome** is essentially the local or regional vegetation, together with all the other living things present within this community. Delimiting different biomes might seem arbitrary, but we do require means of describing different and contrasting communities around the world. Clearly, the vegetation of the Arctic tundra, the Sonoran desert, and the tropical rain forest are all very different. The vegetation that defines these biomes is marked by its stature, density, productivity, and its annual cycle of life activities. Many of the world's biomes also differ greatly in the species numbers they support, each constrained largely by the environmental stresses they experience. Discussions regarding biomes have also been framed in terms of life zones or ecosystems: areas of similar climate and vegetation structure in different parts of the world. Hiking up the side of a tall tropical mountain allows you to experience a variety of "life zones," beginning from a hot and humid rain forest, through cooler montane cloud forests, and ending in treeless alpine meadows. Biomes, or life zones, are a major aspect of the living world, and we will discuss them in chapters 4 and 5. Setting the subject of biodiversity aside for a moment, let's reflect on diversity beyond our own illustrious planet.

DIVERSITY IN THE UNIVERSE AT LARGE

Variety, we've discovered, is almost everywhere we look. Stars vary greatly in size, luminosity, and even color, while galaxies come in a variety of shapes and sizes. A profound revelation during the last half of the twentieth century was finding unexpected diversity within our own solar system, thanks to interplanetary probes. No one had imagined that the larger moons of Jupiter and Saturn might be so different from each other. Europa, a satellite of Jupiter, seems to have a smooth covering of solid ice and might even possess a watery ocean deep within.[7] Another Jovian satellite, Io, has a dramatically colorful surface, marked by frequent and violent volcanic outbursts. Likewise, the moons of Saturn come in great variety. Venus, where science fiction had imagined steaming jungles, proved to have a torrid surface baking under a suffocating atmosphere. Despite imaginative conjectures, subtly changing colors on Mars were not seasonal vegetation changes but dust storms occasionally sweeping across its frigid deserts. Even rocks falling from the sky—meteorites—vary in structure and composition. These many astronomical discoveries were concordant with something everyone already knew: history matters![8]

Each of the major planets and their many satellites have had their own special history, and each is different on that account. Likewise, planet Earth has had its own unique historical trajectory. The *Big Whack hypothesis* claims that a Mars-sized object sideswiped Earth early in our planet's history. This impact added mass to Earth's interior, even as it threw debris high into orbit—debris that would condense to become our lovely **Moon**.[9] Swinging around our planet every twenty-eight days, the Moon constrains Earth's axial gyrations. Our spin axis is tilted at an angle of about twenty-three degrees to the plane in which we orbit round the Sun. This tilt gives us our seasons, recurring each and every year. But without our Moon, Earth's axis might wobble back-and-forth by as much as forty degrees over millions of years. Thanks to a large Moon, our axis ranges over only two degrees. By constraining our *wobble*, the

Moon gives Earth a more stable platform for terrestrial life. On the other hand, and as the Big Whack hypothesis suggests, the solar system was quite chaotic during its early history.

Craters covering the Moon and Mercury document an early period of asteroid and cometary collisions. Astronomers propose a "late heavy bombardment" (4.1 to 3.8 billion years ago), which may have been due to resonance perturbations, as the giant outer planets reordered themselves.[10] This scattering of comets and asteroids throughout the solar system may have been another lucky break in Earth's history, peppering our planet with debris from the farther reaches of the solar system. Because the outer regions of our solar system contain more moisture than the inner, this early influx brought us *lots more water!* Without that precious liquid—in grand abundance—oceans would not encircle our globe, rain would not fall across broad continents, and we, quite simply, would not be here.

Even more intriguing is the conclusion that solar systems experience complex dynamics in their early history and that it's only a matter of good luck that we have the arrangement we do, with rocky planets close to the Sun and large gaseous planets more distant.[11] The discovery of "Hot Jupiters" closely circling other stars made it clear that orbital interactions can result in a wide variety of planetary configurations. Jupiter, in fact, may be another factor in Earth's vibrant biodiversity. By far the largest planet in our solar family, Jupiter circles the Sun in an almost perfectly circular orbit. A more eccentric orbit would have perturbed other planetary orbits and resulted in a very different solar system. Thanks to our largest planet being five times as far from the Sun as we are and having a near-circular orbit, our solar family has been stable for a very long time.[12] In addition, Jupiter and Saturn have served as vacuum cleaners, sweeping up errant asteroids and comets with their gravitation, reducing the likelihood of Earth being bombarded in more recent times.

A VERY FORTUNATE PLANET

> *It remains to be seen whether truly earth-like planets are common-*
> *place or whether a unique series of events during the solar system's*
> *formation gave rise to our very special world.*
> —John Chambers and Jacqueline Mitton[13]

We are here because of many lucky breaks: fortuitous events that have contributed to making our planet so comfortable for life. To begin, Earth circles the Sun in what's been called the "Goldilocks Orbit" (or the **Habitable Zone**). In the children's fable that bears her name, Goldilocks, a very fussy little lady, insisted that her porridge be not too cold, not too hot, *but just right!* Our distance from the Sun keeps Earth's surface not too hot and not too cold—just as Goldilocks demanded. Thanks to a thin blanket of greenhouse gases trapping the Sun's warmth, water remains wet over most of our planet's surface. Just as important, Earth's orbit is near-circular, resulting in seasons that are not as severe as they would be with a more elliptic (eccentric) orbit. Temperate areas do have warm summers and cold winters, but these are caused by Earth's axial tilt—not the orbit's eccentricity.

Getting whacked by the proto-Moon is likely to have been another lucky break, giving us our twenty-three-degree tilt. This tilt has important effects. Even along the equator, the Sun's path across the sky moves southward from July to December, then northward between January and June. Aligned with the Sun, the intertropical convergence zone sweeps seasonal rains back and forth across the tropics each year. This is why tropical monsoons appear with such regularity at the same time each year. Without our axial tilt, tropical rains would fall continuously along only a narrow equatorial band. What Earth's wobble does through the year is to spread rain more widely across the tropics. This same wobble provides long days during the Arctic's short summer, warming tundra vegetation, sustaining forests in Siberia, and allowing wheat production in Alberta.[14] Annual seasonality is a significant factor supporting terrestrial biodiversity, all thanks to a tilted axis.

Earth's size is also *just right!* A larger planet would have much stronger gravitational attraction, making agile terrestrial animals—such as monkeys—impossible. A larger planet would also trap a denser atmosphere, perhaps suffocating the surface under a Venus-like blanket. On the other hand, a smaller planet with weaker gravitation simply can't hold onto a sufficiently dense atmosphere, essential in sustaining active terrestrial life. And there's more. Earth's daily spin means that days do not become too hot, and nights do not become too cold. A faster spin would result in continuous gale-force winds, making the evolution of terrestrial vegetation unlikely. A slower spin would result in long, frigid nights, followed by long, torrid days. Under such conditions, marine life might be able to survive, but life on land would be impossible.

Earth is unique within our solar system in yet another fundamental way: we've got **plate tectonics**! Thanks to a planet rich in water, hydrated minerals—heated at depth—give our planet a dynamic crust. Beginning perhaps three billion years ago, the Earth's crust broke into plate-like sections. With new magma rising along fissures in the ocean floor and spreading laterally, continents are pushed around Earth's surface on giant plates. Slamming into each other, one plate must override the other, subducting the ocean floor while forcing mountains high into the sky. Mountains, as we shall see, are a huge boon to terrestrial biodiversity. Just as important, plate dynamics maintain stable continental platforms well above sea level. Parts of Canada, Australia, and Greenland have geology more than three billion years old. Continuing erosion should have worn down these surfaces a long time ago. Such eroded surfaces would have ended up around eight thousand feet below the sea—all around the globe. Surely, without plate tectonics we'd all be fish!

Plate tectonics has played another critical role as well. Carbon dioxide is absorbed by rain water, producing weak carbonic acid. In this form it reacts with rocks to form carbonates and can be used by marine animals to build their many shells. By removing carbon dioxide from the air, these processes sequester carbon at the bottom of the sea. That's bad news: carbon dioxide will become ever more rarified in the atmosphere and, without this greenhouse gas, the planet's surface might cool into a frozen ball. Enter plate tectonics! In regions where the ocean floor is subducted under advancing plate margins, sea floor sediments are heated at depth and volcanoes form, *belching carbon dioxide back into the sky!*[15] By returning sequestered carbon back into the atmosphere, plate dynamics have been essential in keeping our planet warm and life-sustaining over the last three billion years. Plate tectonics helps maintain continents high and dry, builds lofty mountain chains, and reinjects carbon dioxide into the atmosphere. And there's more: by forming hot cauldrons beneath colliding plate margins, elements and minerals can become concentrated into rich ores. No other member of our planetary family has a similar process and none have concentrated ores in near-surface deposits. ("Mining" the Moon or Mars is fantasy.) Without plate tectonics we'd have little in the way of metallurgy.

The more we learn about our planet and our solar system, the more special they appear to be. Interstellar space is a near vacuum, with temperatures only a few degrees above absolute zero. Take a spacewalk without your spacesuit and you'll be freeze-dried on the spot. In contrast to the frigidity of space, star interiors burn at millions of degrees, even as their outer shells glow at thousands of degrees. In this grand universe, Earth-like habitats with moderate temperatures are very scarce indeed.

Most importantly, our Sun is not an ordinary star. First, the Sun is a *solitary* star, without a costar to bollix-up the regular orbiting of its planetary family. The Sun is also a larger star; more than 80 percent of stars are smaller and give off less energy. Keeping water wet requires that a planet must circle closer to a smaller star, and that's a problem! A closer planet is very likely to become bound to its star in tidal lock. With one side always facing its star, finding terrestrial life forms on planets circling smaller stars seems utterly unlikely. (Earth holds the Moon in tidal lock, which is why we always see the same face of our splendid satellite. Mercury and Venus are both in tight resonance with the Sun, revolving only very slowly.[16])

Another pleasant astronomical detail is that because the Sun is a larger star it is more stable than many smaller stars. Our Sun came into being about 4,560 million years ago. A single solar hiccup could have put an end to life on Earth over these last few thousand million years. Lesser stars are not as uneventful; they flare up from time to time. In addition, our star orbits the crowded and explosive center of the Milky Way Galaxy at a safe distance. Who knows what havoc a nearby supernova explosion might wreak upon a biosphere? More than four thousand million years of calm stability have resulted in an ongoing scenario of ever-more complex life on planet Earth. Viewing ourselves from a broader astronomical perspective, we are the lucky inhabitants of a very rare and special place.

Many have claimed that, with billions of likely Earth-like planets, civilizations like ours must be common in our galaxy. However, the more we learn, the more unlikely that appears. SETI—the Search for Extraterrestrial Intelligence program—has been scanning the firmament for radio signals over more than forty years, and they have failed to intercept a single coherent message.[17] More fundamentally, complex biological beings did not evolve to traverse the vastness of interstellar space; if interstellar travelers exist they'll be robots capable of "sleeping" over many thousands of years. Remember that stars are separated in distances measured by "light years" and, with light speed at 186,000 miles (300,000 km) *in a single second*, interstellar travel by living things remains a fantasy. Setting astrobiology ("the science without a subject") aside we'll confine ourselves to the planet on which we live.[18]

OUR EVOLUTIONARY EPIC

Over the last five centuries, science has fashioned a grand history for our species, our planet, and the universe itself. Bringing all this information together has given us a coherent narrative across time. We call it the **Evolutionary Epic**. This scientific effort resembles earlier mythical epics, fashioned by cultures around the world, as people tried to make sense of the world around them. Though deficient in purpose and devoid of meaning, our scientific epic is extravagantly rich in detail, while ranging over enormous scales in both time and space. From within the nucleus of an atom to galaxies being flung ever farther apart, science has revealed a cosmos larger and more complex than anything imagined earlier.[19]

Though scientists may delight in so grand a historical narrative, many people find it unsatisfying. Lacking an explanation for the universe's origin, with little linkage to revered sacred texts, and seemingly without purpose, a great majority of people continue to find comfort in traditional religious narratives. But science is different; it is nothing more than a pragmatic way of trying to understand the world around us. While many scientists examine the real world through carefully controlled experiments, the origin and elaboration of biodiversity are historical questions. In these instances, we formulate historical scenarios and then seek evidence from nature to support or reject a given scenario. It's very much like detectives trying to solve a crime. In the late 1800s, Thomas Huxley hypothesized that humans originated in Africa for the simple reason that our closest relatives—chimpanzees and gorillas—live there. After more than a hundred years of effort, we now have evidence supporting Huxley's original conjecture. Fossils in the ground, as well as DNA within our bodies, indicate we humans originated in Africa.

Religious fundamentalists have responded to the evolutionary epic by claiming that the complexity of many biological systems can only be explained as having been "designed," implying the direct intervention of God. While that's fine for most people, for science there's a problem. How can this general hypothesis be tested or evaluated? Furthermore, if God has used magical or miraculous means to fashion the universe, there is simply no way that we humans could comprehend the universe; we're no good at magic. On the other hand, if the "Creator" used the natural laws we see around us in creating our world, then science can hope to understand the world. If God sort of kick-started the universe and then gave it a set of rules and regulations (the strength of gravity, the speed of light, the strong nuclear force, natural selection, etc.) then we should be able to figure things out.[20] Following this strategy, scientists of all faiths have found common ground in studying the world in which we live.

Science, as Nicholas Wade points out, "consists largely of facts, laws and theories. The facts are the facts, the laws summarize the regularities in the facts, and the theories explain the laws."[21] Evolution, whether of the planet or life forms, is paleontological fact. In contrast, **Evolution by Natural Selection** is a theory that helps explain the "design" we find in nature. Together, evolutionary fact and evolutionary theory have given us a single coherent narrative. And why not? There is only one real world out there! Meteoritic studies (using elemental isotope decay rates) indicate that the solar system is about 4,560 million years old. This long history has provided a wealth of time for life to develop and proliferate. Paleontology gives us a picture of Earth history that coincides with times determined by isotope decay ratios within cooled and solid lava. In addition, fossil-bearing sedimentary rocks record how life has changed over time, expanded grandly, then fallen back several times, only to recover and form even more diverse biotas. Indeed, the geological time scale is linear and consistent around the planet. Mastodons and humankind coexisted during the recent ice ages; neither was present when dinosaurs ruled the world. Similarly, the great coal measures of the Carboniferous period had roaches, dragonflies, and a few amphibians, but neither dinosaurs nor mammals occur in these more ancient deposits. Biblical fundamentalists claim that an enormous flood carved out Arizona's Grand Canyon, but they cannot explain thousands of feet of canyon walls, successively layered with geological strata. Nor can they explain why the lowermost strata at the Grand Canyon are completely devoid of animal fossils, while the uppermost contain dinosaur fragments. Today there is no question: our planet's long history has provided the time, the lucky breaks, and the many opportunities—along with a few calamities—to produce the exuberance of living things we see today.

BIODIVERSITY: ON LAND AND IN THE SEA

A surprising fact regarding today's biodiversity is that the numbers of species of plants and animals living on land far outnumber species living in the oceans.[22] Estimating the world's habitable volume, from the deep ocean to the highest mountains, we find that over 90 percent of that volume lies within the sea. Why might such a large volume have fewer species? The oceans do have a great diversity of animal phyla, from jellyfish and a variety of worms to crustaceans, mollusks, fish, and plant-like bryozoans. Life began within the sea and, obviously, that is where the greatest number of ancient plant and animal lineages are to be found. Green algae, brown algae, red algae, diatoms, and many microscopic forms are all abundant in the ocean. These lineages represent divisions of great age. This is diversity at the highest ranks: animal phyla and plant divisions. However, diversity at the lowest rank—species numbers—presents a completely different picture. Insects alone number around 800,000 described species, far outnumbering the 250,000 described plant and animal species in all the oceans! Marine biologists contend that there are many unrecognized species in the sea, and that may be correct. However, we needn't worry ourselves about such marine life, mostly microscopic and still waiting to be described. Instead, this book will concentrate on the numbers of larger terrestrial creatures.

Though land areas amount to only about 29 percent of the Earth's surface, scientists estimate that a bit more than half of the world's photosynthesis takes place on land.[23] Photosynthesis by green plants, various algae, and blue-green bacteria captures the energy of sunlight to build energy-rich carbohydrates and other foods. Whether on the land or underwater, photosynthesis keeps 99 percent of the living world from running out of gas.[24] Energy is the fundamental currency of life. It makes no difference if you're a bacterium or a ballerina, when you run out of energy your life has ended. Despite the huge volume and large surface areas of ocean water, the ability of photosynthesis to capture sunlight and absorb carbon dioxide is limited in marine environments. There is insufficient light for photosynthesis in deep water, and many areas of the ocean are poor in nutrients. Worst of all, there are hordes of animals, large and small, ready to devour those who photosynthesize within the sea. Also, a vast majority of marine photosynthesizers are microscopic and short-lived; sea grasses, larger kelp, and other algae are uncommon. Neither the Cyanobacteria nor microscopic algae can provide the physical and nutritional support afforded by trees, shrubs, and grasses. All told, esti-mates of biomass (the weight of living matter) in the oceans range from five to ten petagrams (5 to 10 billion metric tons). This contrasts with around 560 petagrams for biomass on land. That's perhaps an eight-fold difference, and it is a major reason for greater biological species richness on land.[25] What's more, Vaclav Smil estimates that standing phytomass (plant biomass) on land may be as much as two hundred times greater than phytomass within the sea.[26]

Biodiversity on land, however, is not simply a product of greater pho-tosynthetic potential. Terrestrial habitats vary from frigid polar regions and parched deserts to evergreen tropical rain forests. They range from mangroves along tropical sea coasts, sub-desert thorn bushes, a grand variety of forests, to mountain-top grasslands. Terrestrial ecosystems are both hugely diverse and they differ profoundly in how stressful they are for their living occupants. You won't ever dry out if you spend your life living in the sea, but desiccation is a serious problem in most terrestrial environments. Also, temperatures vary over only a modest range within the world's oceans. Marine environments offer nothing comparable to the stress of living on a Midwestern prairie through the year, where win-tertime temperatures can plunge to −30° C (−20° F) and summertime temperatures can reach 40° C (102° F). Clearly, the challenges of living on land are far more varied than those within the sea. Terrestrial substrates are diverse as well, ranging from rich loams and deep mud to clay, loose sand, and bare rock. In fact, the soil itself may house thousands of bacte-rial species in a single gram.[27]

Another factor promoting terrestrial biodiversity is that many land surfaces are separated by broad oceans. Such isolation has resulted in dis-tinctive land biotas on different islands and distant continents. Fish can be very diverse around a coral reef, but there is the potential of sending their little ones off to join another coral community a thousand miles away. In contrast, few rain forest inhabitants can journey across a hostile ocean. The Congo rain forest may look like an Amazonian rain forest or one in Asia, but a closer look will reveal the many different species and genera comprising these widely distant biomes. A terrestrial world, divided by oceans large and small, and rich in environmental variety, has produced an astounding number of plant and animal species.[28]

Not only is there isolation between land masses separated by the sea, there can be isolation within the larger land areas themselves. If you are adapted to living on a mountain top, your offspring might not be able to reach another, distant, mountain top. Likewise, if you are a freshwater fish living in a river on the west side of a mountain range, it is unlikely your offspring will cross that range to swim in rivers on the eastern side. Of the approximately 31,000 species of fish gracing our planet, a full third live within fresh water, even though fresh water makes up only 1 percent of the water on our planet. The Amazon basin is estimated to have as many species of fish as live in the entire Atlantic Ocean! With so much diversity on land surfaces, and since we are creatures of the land ourselves, we will focus on terrestrial biodiversity and our own history.

This book celebrates a simple fact: our planet has become increasingly complex over time, whether with geological dynamics, continually evolving life forms, or new human technologies. We will explore the many factors that have propelled this extraordinary epic forward over the last four thousand million years.

To better appreciate our planet's biological wealth we'll address a number of questions. To begin, we'll examine the insects and try to understand what has made them so successful. In chapter 2, we'll ask how bacteria fit into discussions of biodiversity, and why larger, nucleated cells were so important an advance in the history of life. For many readers this may be a bit of a slog, but living things are complex and we need to understand basic processes in our quest to understand life's forward momentum. In chapter 3, we will discuss the factors actually generating new species.[29] Chapters 4 and 5 will survey species numbers around the globe, look for general patterns, and try and understand why some regions harbor unusually high numbers of distinctive biotas. In chapter 6 we'll tackle the question of how so many species manage to live together in the same biome at the same time.

Shifting from contemporary ecology, we'll turn to paleontology in chapter 7, examining the history of biodiversity and increasing complexity over the last 600 million years; in chapter 8 we will examine the factors driving these ongoing trends. In chapter 9 we'll discuss the origin of what we like to think of as nature's most complex organ: the human "mind." With this device, and after forming effective symbioses with a few plants and animals to better feed ourselves, the stage was set for explosive cultural advance. Chapter 10 will review how settled human communities increased their technological prowess, and chapter 11 will review the four billion years that have led to human dominion. Our last chapter will examine how expanding technologies and our increasing numbers have become so transformative for the biosphere—and so problematic. Beginning our journey, let's start with a really successful crowd: beetles, the six-legged kind.

Chapter 1

BILLIONS OF BEETLES

There is a bit of old gossip that continues to be retold amongst biologists. According to this tale, the eminent British biologist, J. B. S. Haldane, was attending a cocktail party when he was confronted by a prelate with a rather unusual question. Repeated ever since the late 1930s, this brief encounter has become a staple when discussing the diversity of life on our planet. The religious gentleman's question was: "As a student of biology, Dr. Haldane, what can you tell us about the nature of God?" Unfazed by the gravity of the question, the eminent biologist is said to have responded simply and directly: "An inordinate fondness for beetles!"

Yes indeed! The Creator seems to have a fondness for beetles. Why else would there be so many of them? At latest count, beetles number around 380,000 described species.[1] What this means is that a student of insects has gone to the trouble of providing the description for a new species, given it a Latin name, designated the specimens to which the name applies, and published all this in a scientific text or periodical. Nowadays, students of insects (entomologists) like to have an illustration published along with the description of a new species, and they insist that a designated specimen—the *type specimen*—be deposited in a public museum or a university collection.[2] All this effort, all over the world, has given us around 380,000 scientific names for beetles! Compare that with about 5,500 species of described living mammals, or a bit more than 10,000 species of birds. Shifting to the plant world, there are over 310,000 described species of land plants on our resplendent planet. This is a conservative number, including mosses, liverworts, ferns and their allies, all the conifers, as well as more than a quarter million flowering plants. Add all these green plants together and they still don't reach a number of species that can match the beetles. Returning to animals, there seem to be more species of beetles than there are species of all the non-microscopic animals living in all the oceans. This aquatic crowd includes jellyfish, many kinds of worms, mollusks of all sorts, lots of crabs, and thousands of fish species. *Three-hundred and eighty thousand described species are a whole lot of species!* And, as if that weren't trouble enough, their numbers keep growing as scientists discover new and undescribed species living high in rainforest trees, and even little ones hiding in the soil not far from where many of us live.

WHY ARE THERE SO MANY KINDS OF BEETLES?

Beyond God's pleasure, what might be the reasons for beetles being so numerous? Like so many puzzles in biology, this question elicits many answers. Surely, the first of many factors is the most obvious: beetles are little. The largest weigh about as much as a mouse, the smallest less than a fly. In strong contrast: there are fewer than a dozen species of rhinos and elephants alive in the world today. These are the largest and heaviest land animals, each requiring a range of many square kilometers in which to live and prosper. Beetles, on the other hand, can live their short lives within only a few square meters. Here we have a pattern seen throughout the living world: numbers of both individuals and species decline as the sizes of animals increase. Clearly, being small makes it possible to pack a lot of beetles into the same small area.[3]

Small size is just one factor helping us understand the diversity of beetles. Beetles are found in many diverse habitats, carrying on a grand variety of life activities. Huge numbers live in both temperate and tropical regions; a few can be found in the tundra and on high mountains, but none are resident in Antarctica. We know how flour beetles can invade our kitchens, carpet beetles damage wool rugs, and Japanese beetles chew on garden plants. In a warm June evening, june bugs (really beetles) fly into our screens, while fireflies (also beetles) flash signals in the dark of early evening. In addition to a grand number of species, some beetle species have populations made up of millions of individuals.

In the American West, ladybird beetles (also called ladybugs) move high into the mountainsides and congregate together to spend the winter. During the early 1900s, collectors would go into the Sierra Madre highlands of California to collect these slumbering beetles, selling them to farmers in their battle against aphids. (Ladybird beetles, both larvae and adults, feed on aphids.) This collecting activity, in the early part of the last century, saw the gathering of several tons—that's right *tons*—of beetles each winter. When you calculate that there may be 44,344,000 ladybird beetles in a ton of 'em, you've got a real lot of beetles.[4] The take-home message is clear: by combining 380,000 *species* of beetles, with *populations* that may reach millions of individuals, it is no exaggeration to claim that our planet does indeed support billions of beetles. But what is it about beetles that have made them such a grand success?

Extinction is a constant erosive force in the history of life. A long fossil record tells us that perhaps 98 percent of all the species that ever lived are already long gone. Surviving changing climates, droughts, disease pandemics, and the occasional geological upheaval is a challenge a great majority of plants and animals have failed. Thus, when we see a lineage as successful as the beetles, we know they've got to be a well-adapted crowd. In fact, a simple glance at most any beetle will reveal their primary survival attribute: they are tough. Trekking over the leafy forest floor you can step on nearly any beetle, only to have it scurry away as you lift your boot. A hard exoskeleton amply protects a delicate interior! The key innovation here is that what was once a forward pair of wings became modified into hard wing covers (elytra) protecting the top of the entire abdomen.[5] Their front and bottom sides are also armor encased. In fact, many species look like miniature tanks. The "Volkswagen beetle" got its moniker because this automobile really did resemble some beetles. Best of all, the tough wing covers of beetles can be raised up, allowing the actual wings to unfold and fly into the distance. Being rather solid means that, while most beetles can fly, they are neither fast nor agile. And that's why the larger ones fly only at night, avoiding the pursuit of predatory birds in the light of day. Being well protected, and with the ability to fly, gives them two significant advantages in the struggle for survival. But beetles have another trait that is even more significant.

A DIVIDED LIFE-HISTORY

Beetles belong to a large assemblage of insects with a four-stage life cycle; they undergo a complete metamorphosis (and are called holometabolous). Such insects have their lives divided into four separate and distinct stages. The first is the egg, which may be a short-lived stage, or it may be a means of surviving a harsh winter or a long dry season. At the appropriate time, the egg is broken open by its little occupant, a small larva. The larval stage is a significant part of the life cycle, being devoted entirely to eating and growing. Beetle larvae range from minute worm-like beings to the larger grubs we often find in soil and rotten wood, while others lead an active predatory life. Whether sedentary or active, this part of the life cycle is devoted to accumulating the energy needed to build the adult insect.

Becoming transformed from a fat grub to an elegant adult requires a separate stage: the pupa. While you may be more familiar with the pupae (plural of pupa) of butterflies and moths, all insects with a four-stage life cycle have this stage. It is within the pupal skin that much of the insect dissolves! That's correct: almost everything within their outer skin becomes liquefied. This fluid, reorganized by several special regions within the pupa, then develops into the fully adult beetle. A miraculous transformation, when you think about it, this is a significant factor in the success of many insect species. The adult can develop into a form completely distinct from its larva, able to fly off to find a mate and lead an entirely new and different life. In effect, beetles have two lives: a larval eating phase, and a travelling/reproductive mature phase. That's the good news. The bad news is that this life trajectory comes with a serious limitation: the adult can grow no more! The new adult is as big as it will ever become.

SCARABS AND BURYING BEETLES

Another significant aspect of beetle diversity is their many differing lifestyles. This allows different species of beetles to live in the same environment without getting in each other's way. For example, scarab or dung beetles spend much of their time looking for fresh dung. When they find it, a pair of beetles—mom and dad—will energetically create a spherical dung ball. The ball is rolled to a convenient site for burial, where the ball will house and nourish one of their larvae. Scarab beetles aren't likely to get in the way of burying beetles who busy themselves looking for the bodies of dead mice and birds. Burying beetles, like their scarab relatives, have complex behavioral traits to live their special lifestyles—all geared to rearing another generation of beetles.

As a boy and a collector of almost anything biological, I chanced upon a dead mouse in the woods bordering our country cabin. Here, I figured, was a chance to get a little mouse skeleton for my collections. So I tied the mouse carcass to a nearby stem with thin wire, making sure that no scavenging animal would cart away my prize. I then covered the body with leaves to further protect it. Busy ants, I thought, would clean the carcass and leave me with a fine little skeleton. Coming back only a day later to check on the deceased, I found that my mouse had disappeared. It was no longer under the covering of leaves. But the metal wire was in place, and it led into the ground. *My mouse had been buried!* Sure enough, after watching carefully, I noticed a pair of burying beetles digging to lower the dead mouse even deeper into the soil. As in the case of the scarab beetles, the buried mouse would provide nourishment for a new generation of beetles.

The burying beetle pair will continue feeding on their carcass and watching their young develop. They even use antibiotics to reduce bacterial activity on the carcasses they feed upon.[6] In contrast, scarabs lay only one egg within the dung ball, bury it, and then go on to fashion more balls for more offspring. That's why the African savanna isn't covered with dung, despite herds of zebras and many species of antelope. In fact, Africa is home to over two thousand species of scarab beetles. By burying huge amounts of dung, these beetles perform important ecological services. Dung (cow patties are an example) can depress the growth of plants beneath them, serve as nourishment for hordes of flies, and help spread parasites and disease. In addition to removing dung from the soil surface, busy scarabs effectively fertilize the soil and help aerate compacted soil.

The importance of dung beetles became clearly apparent in Australia. Cattle were brought to Australia to develop a beef industry—great for producing hamburgers, but not so good for the landscape. It turned out there were no larger dung beetles in Australia to deal with the cow pies produced by our large domestic cattle. Soon pasture lands were bedecked with bone-dry, rock-hard cow pies. This new groundcover just sat there, like so many concrete platters, reducing the absorption of sparse rain into the soil, suppressing the growth of grass and forbs, and dramatically reducing the productivity of the land. Clearly, bigger species of scarab beetles were the missing element in the Australian outback.

Beetle variety is awesome, and not just in numbers. Diving beetles are voracious little predators in lakes and ponds, both in their larval and adult phases. Since they devour aphids, ladybird beetles are welcome in our gardens. Fireflies, or lightning bugs, fascinate us with their ability to produce flashes of yellow-green light. But there are multitudes of other kinds of insects as well. Dangerous wasps with potent poison and the stingers to deliver it; elegant dragonflies along the shores of lakes and streams; smelly stink bugs; bumblebees working one flower after another; the pesky flies, and lots more. Let's move beyond beetles and consider all the insects.

THE NUMBERS OF INSECTS

As we search for additional lineages with huge numbers of species, we find that these are also insects. The flies, order Diptera, number about 150,000 species, while butterflies and moths (Lepidoptera) have over 120,000 species. Put these all together, add in the beetles and lesser groups, and we probably have around one million species of described insects. And that's just at our current state of knowledge. Ants, bees, and wasps (the order Hymenoptera) currently number around 115,000 named species. However, one expert thinks the real number inhabiting our planet is closer to a million![7]

These are huge species numbers in the world of living things. In addition, some species include large numbers of individuals. Surely one of the most spectacular expressions of insect numbers is that of locust swarms. Not only impressive in their vast numbers, these far-flying grasshoppers can devastate agricultural crops, threatening people with starvation. The most notorious of these is the desert locust (*Schistocera gregaria*), ranging from northern Africa to India. These are the creatures mentioned with foreboding in the Bible. My students and I witnessed a swarm of these migrating pests in eastern Ethiopia during a February field trip. Having stopped along a high mountain road, we looked into a distant valley and noticed an odd "cloud." It was morning, and I assumed it was persisting early fog. But it did look strange; the moisture of early morning should have been long gone. The next day that very same cloud—a horde of migratory locusts—flew over our college campus. Probably about two kilometers long and half a kilometer wide (1.2 by 0.3 miles), this cloud took almost an hour to stream by. Standing under the passing horde, I couldn't begin to comprehend the numbers passing overhead. Fortunately, they weren't stopping to eat; they were in full travel mode. A few days later we saw this same swarm flying over the region's highest mountain, Gara Mulatta. To get across that mountain, these locusts were flying at an altitude of 3,400 meters (11,000 ft.)! Estimating the numbers of a particular swarm in East Africa, researchers came up with a total figure of about fifty billion individuals. And with each locust weighing around

2.7 grams (a tenth of an ounce), that swarm added up to 115,100 tons of locusts. That's a lot of grasshoppers! Nowadays, with global weather satellites showing rainfall patterns, locust outbreaks can be predicted, and measures can be taken to arrest their reproduction.

Another insect outbreak of extraordinary numbers occurs exclusively in the eastern United States. Only, in this case, the explosion of numbers is both predictable and long-delayed. Once every seventeen years (thirteen years in southern broods), our eastern woodlands are assaulted by a cacophony of deafening decibels. These are the calls of periodical cicadas that lived their long lives underground, sucking sap from the roots of trees and shrubs. Lacking the four-stage life cycle of holometabolous insects, cicadas—like grasshoppers—grow through a series of nymphal stages. Emerging together in the late spring of their seventeenth year, the nymphs crawl out of the ground and up stems and trunks. Once securely positioned, their skin splits open along the back, and the soft-bodied adult emerges. Hanging quietly for a few hours, wings expand and their exoskeleton hardens; they are now ready to fly. Soon the males begin their ear-splitting choruses; hoping to attract the ladies. Indeed, hiking through a forest harboring thousands of these tree-top choristers can be very uncomfortable! (Fortunately, they stop calling at nightfall.) Three different species of cicadas make up most of these singing swarms, and they are divided into a number of distinct geographical broods, each emerging on its own seventeen-year or thirteen-year schedule. While cicada species are found throughout the tropical and temperate world, only these particular species in the eastern United States display a uniquely synchronous emergence.

A DISTINCTIVE BODY-PLAN

Visually, the most obvious characteristic of insects is their legs. Virtually all mature insects have six legs. It makes no difference if you are looking at a fly, a bee, a grasshopper, an ant, a wasp, a termite, a roach; they've all got six legs (and are called hexapods). Here we're talking about adults; some caterpillars have six smaller legs up front and pairs of fat legs along the back segments. And then there are many insect larvae that have no legs at all (think of maggots), but for adults the picture is pretty uniform: six legs. In some, like the praying mantis, the front legs are adapted for grabbing prey, but it's pretty clear these are modified legs. Insect legs are jointed with three main sections, allowing for a wide range of movements. Lastly, claws at the end of the legs allow for grasping. Jointed legs give insects flexibility in posture, in movement, and six attachment points for hanging tight.

Another distinctive character is having an adult body divided into three sections: head, thorax, and abdomen. In fact the Latin word *insectum* means division. The **head**, up front where heads are usually found, holds the primary sensory organs (antennae and eyes), as well as the feeding apparatus. Mouthparts differ widely in different insect lineages, from those with plier-like pincers to those with piercing-and-sucking mouthparts. Having a grand variety of feeding styles allows different lineages to chow down a great variety of food sources. The second part of the body is called the **thorax** and provides a solid central section where legs and wings are attached. Here is where the muscles for flight and locomotion are located. The last part of the body is the **abdomen**, housing most of the intestines, breathing apparatus, and reproductive organs. All of these sections have a covering of rigid chitin, giving the body both rigidity and protection. But what might have been the primary key to insect success?

A GRAND INNOVATION: FLIGHT

Flight is perhaps the most significant of insect attributes. The fossil record indicates that insects were the first animals to propel themselves through thin air. Giant dragonflies flew through coal-age forests around 300 million years ago, and all the larger lineages of insects are capable of flying. The wings are not modified legs, and it isn't clear how insect ancestors managed to evolve two pairs of wings. By being attached to the top of the thorax these wings remain clear of the legs below and allow for mobility in flight.

A further advance was the ability to fold their wings against the body, great for hiding oneself in a safe niche. Two ancient lineages, the mayflies and dragonflies, cannot fold their wings against their bodies, and cannot tuck themselves into tight corners. Flight, in fact, marks the full maturity of insects. When fully developed, the wings cannot be replaced; once damaged they stay damaged. Whether having a gradual development (like grasshoppers, cicadas, and true bugs) or with a four-stage life cycle (like beetles, flies, and wasps) flight marks the end of further growth.

Have you tried to swat a fly? Not an easy task—unless you've got a fly-swatter. Highly sensitive, house flies can feel an increase in air pressure as your hand moves toward them—and off they go. Overall, flies (order Diptera) are characterized by having only a single pair of wings. Behind the wings, flies have two short, slender, club-like devices called halteres. These vibrate during flight, helping maintain stability and orientation. Unfortunately, the flies include some of the nastiest organisms on the planet: mosquitoes, biting flies, and a variety of winged parasites.

The advantage of flight, coupled with a four-stage life cycle, is epitomized by nasty flies. Think of a female mosquito flying far and wide, seeking a blood meal that will allow her to create a raft of eggs. Placed on stagnant water, the eggs will give rise to predatory aquatic larvae (no lazy grubs these). The pupae float at the water's surface to release an adult mosquito capable of flying off to seek a mate, get a blood meal (in the case of females), and then finding another body of water to continue the life cycle. There are even more intricate strategies among flies.

While on a collecting trip in Costa Rica's wet and soggy Tortuguero National Park, I had forgotten my insect repellent and was pounced upon by a good number of mosquitoes. Not to worry, malaria wasn't prevalent in this area, and I figured the itchy mosquito bites would subside in a few hours. Not so! Two weeks later and back in Chicago, I still had three itchy bumps that kept right on itching. Worse yet, they were getting bigger! These things looked like growing tumors, and that was scary. So, off to a dermatologist for a closer look. Not recognizing what he was looking at, he scheduled a biopsy for the following week. But before the scheduled biopsy, while waiting to go to sleep, I was making myself miserable imagining cancer growing under my epidermis. The three enlarging tumors were located on my right thigh, my left arm, and at the top edge of my forehead. Then, quite suddenly, while lying there in bed, the "tumor" on my forehead went **skritch, skritch, skritch**!

This "tumor" was making a noise! Now, I knew almost nothing about cancerous tumors, but I did know they do not make audible sounds. Remembering a similar incident in Ethiopia, I realized there were three botfly larvae under my skin! Next morning, I gleefully phoned the dermatologist, and he quickly scheduled simple outpatient surgery to remove these beasts. Arriving at the hospital's surgery room, I was greeted by a crowd: doctors, nurses, and a bunch of medical students. In Chicago, there weren't a lot of botfly maggots to be dug out from under people's skin. Nevertheless, this was a show I didn't want to see, so—lying on the operating table—I stared quietly at the ceiling. Surgery proceeded and, as the second maggot was being excavated, a tall surgeon rushed into the room still wearing all his operating-room attire. Looking over the shoulder of the operating surgeon as the last maggot was removed, he blurted out, "This is really gross!"

Not just gross, but fascinating. My body had built tissue around these nasty maggots, so they were confined to my skin. That meant they grew slowly by scrapping away the surrounding tissue: **skritch, skritch**. I still carry a dent in my forehead where the little beast did its skritching. But here's the really extraordinary part: *I had gotten these botfly larvae from mosquitoes!* It seems that botflies aren't very good at finding warm mammalian bodies; but mosquitoes are! What the female botfly does is to grab hold of a female mosquito and attach an egg to one of the mosquito's legs; the egg then drops off the mosquito when the latter makes landfall on a warm mammal. This sounds more like science fiction than natural history, but it is a great example of how insects have been able to forge unique life histories, coupling flight with complex behaviors. A Field Museum zoologist recalled having collected a puma specimen many decades earlier in southern Mexico and finding many botfly larvae as they skinned the big cat.

Many genera of wasps are also flying parasites. Check out your tomato plants in late summer and you can often find a tomato hornworm festooned with little white cocoons over its body. These are made by minute parasitic wasp larvae that have finished feeding and are pupating on the doomed caterpillar. Here, natural selection has fashioned a very clever strategy. Such maggots feed first on non-vital parts of their host, before devouring the vital organs. In fact, the world's smallest insects are parasitic wasps. Inserting one of their eggs into the egg of another insect species, the little larva feeds within the solitary egg, finally emerging as a brand new little wasp.

All told, with their ability to fly, a multistage lifestyle for many lineages, and their small size, insects have proven to be one of Mother Nature's most successful life forms.[8] Leaf miners, the caterpillars of very small moths and flies, do all their feeding between the top and bottom surfaces of a single green leaf. The world's smallest beetles live in decaying leaf litter on the forest floor. About a millimeter (1/24th of an inch) in length, these featherwing beetles have gotten as small as is possible for a beetle to be. The female featherwing can produce only one egg at a time; a beetle can't get any smaller than that!

Returning to the huge numbers of beetle species, I should mention the work of Terry Erwin in Panama. By fogging trees of the same species (*Luehea seemannii*) with insecticide, Erwin collected all the fallen insects on stretched canvas beneath the trees. In this way he attempted to survey the numbers of canopy insects in a single species of tree. Having fogged nineteen trees, he collected a huge number of little creatures, including almost 8,000 beetle specimens, representing about 1,200 species. That's just one tree species, but Panama is home to around 2,000 different species of trees! If some of these beetle species are found in only one or two tree species, you can imagine the numbers of canopy beetles in Panama.[9] And surely, without the ability to fly, these many beetles could not maintain their high-canopy lifestyle. But despite their many forms and huge numbers, this six-legged alliance has not been able to get really big or really smart.

WHY HAVEN'T INSECTS GOTTEN BIGGER OR BRAINIER?

Each larger grouping of animals is tightly confined within an invariant body plan. Insects have six legs, spiders have eight legs, and crabs have ten. Our own lineage is similarly constrained. The land vertebrates, from frogs and toads, through reptiles, birds, and mammals all have four legs. Birds changed their forelegs into wings, and we changed ours into arms, but that's just a variation on an underlying and unchanging theme. Our lineage arose from early four-legged amphibians, and we are stuck in that format as tetrapods. It seems that once a new developmental protocol, such as a three-parted body plan in insects, becomes established, it is virtually impossible to change. Loss, of course is another matter; loss is easy and has occurred often. Many insects have lost their wings. Snakes have lost their legs, while dolphins transformed theirs into flippers; but the basic body plan, what German-speaking zoologists called *Bauplan*, remains invariant. And that's what makes it so easy to recognize crabs, spiders, and insects as crabs, spiders, and insects. But mammals range from little shrews, weighing less than a goliath beetle, to elephants and whales. Why haven't the insects done the same?

The reasons may be several. For one thing, a hard exoskeleton needs to be shed to expand body size. Grasshoppers (in the order Odonata) and true bugs (in the order Hemiptera) go through a series of molts before they are mature. These insects do not have the four-stage life cycle we've talked about in beetles and flies. Their eggs produce a little nymph, which proceeds through larger stages, molting at least four times to become the mature reproductive adult. Young grasshoppers, in fact, look like wing-less adults. In contrast, the holometabolous insects with four life stages, confine all their growth to the larval stage and follow this with a major transformation within the pupa. Think about the tomato hornworm caterpillar as its mouthparts devour the leaves of your tomato plant. Fully grown, the larva will pupate and will give rise to a hawk moth, with a curled-up "tongue" that can be extended to four inches long (10 cm) and gather nectar from deep-throated flowers, while hovering with rapid wingbeats (like a hummingbird). From tough chewing mouthparts in the caterpillar to a long slender tongue in the moth, the same species lives two very different lives.

Pupation has given the holometabolous insects a platform from which to explore many different lifestyles. However, all insects remain constricted when reaching full maturity. Whether grasshopper, beetle, wasp, or butterfly, the mature insect cannot become larger because it can no longer shed its skin nor replace its wings. Really big insects, like the Goliath beetles of African rain forests and the stag beetles of Europe, become so large because they have big grubs. And since most grubs are incapable of repelling attackers, they need to be hidden away in safe surroundings. The larvae of these large beetles feed deep within rotting wood, where they are hard to find. These same large grubs may feed for four to eight years before they reach full maturity and are ready to pupate. That's a long time, and a good reason for staying small, but there's more.

A larger active animal, like a bigger more powerful car, requires more fuel. Not only more fuel, but fuel that has to be burned faster. For animals, that means bringing in a lot of oxygen to keep the fires of respiration going. Surely the best set of lungs among land animals are those of birds. In birds, air passes through the lungs and out in one continuous path; a necessity for fast continuous flight.[10] And this is where insects have a major problem: they don't have anything equivalent to lungs. Air diffuses into the insect through apertures in their sides (spiracles); oxygen is absorbed into their bloodstream, and then pumped around the body. There is nothing comparable to the elaborate alveolae in our lungs, providing large surfaces over which oxygen can be absorbed by the bloodstream. Nor do they have a diaphragm that pumps air in and out of their bodies. Paleontologists conjecture that the giant insects of the Carboniferous period lived during a short period when oxygen became as much as 30 percent of the atmosphere, in contrast to the 21 percent we live with. When oxygen levels returned to lower levels, the giant insects vanished.[11] Because brain tissue requires lots of oxygen to do its work, limited oxygen uptake results in little brains for little insects. The combination of a tight exoskeleton and an inefficient oxygen-uptake system has doomed insects to remaining forever small but nimble.

This presents us with a larger question: why are the insects *stuck* in the body plan that defines them? Similarly, why couldn't land vertebrates, over more than 300 million years, grow themselves another pair of legs? (Imagine being able to scratch your back, while your two arms and two legs are otherwise engaged). It seems that all the larger groupings of animals are tightly constrained within the confines of a stereotypical *Bauplan*. Thanks to recent DNA studies, we now understand why each lineage is confined to a specific body type. Research has shown how a careful program of gene activity transforms a little mass of undifferentiated cells into a much larger and more complex animal. The little ball of cells—the proembryo—comes directly from the fertilized egg, through a series of simple cell divisions. But then developmental protocols kick in and begin to chart a more complex course. They instruct the little group of cells to form a hollow ball, then have part of the wall bend inward, resulting in two cell layers. With expansion and elongation, a head end and a tail end, the bottom and the top, become clearly defined. That's how all us animals begin, whether a worm, an insect, or a vertebrate. These developmental protocols are the genesis of a different kind of diversity: the diversity of cells and tissues within a single complex animal. Once these protocols were established within a lineage, changing them has been all but impossible.[12]

UNDERSTANDING ANIMAL DEVELOPMENT

Robert Macarthur, late ecologist, suggested that biological scientists come in two varieties. The engineers, he proposed, want to know how things work, while the second group, historians, want to find out how things got the way they are. Animal diversity, whether billions of beetles or hundreds of cell types, can be viewed from the same two points of view. First is the here and now of how things work. So many beetles living today; so many different tissues making up the body of a single fly; so many complex processes in each living being. In contrast, there are the historical questions: How is it that our planet got to have so many beetles? And how were the developmental protocols that produce such a variety of plants and animals initiated?

Both the development of a human baby from simple embryonic beginnings and the emergence of a butterfly from its pupal skin are awesome transformations. Each is a grand advance in structural complexity, and both are driven by similar genetic protocols—though each is unique to its particular species. The fact that such developmental trajectories have produced billions of beetles should not diminish our astonishment. Some things we see around us are truly wondrous, even when there are a lot of them.

Today, with a rapidly expanding understanding of the genetic code, and the realization that fruit flies, mice, and people all use similar genes to build their bodies, the glory of morphological development is yielding its secrets. No longer the product of mysterious miracles, the similarity of genetic instructions in a wide array of animals has made clear how Mother Nature has used common tools, over and over again, to build the huge array of living beings that call our planet home.

WHY ARE THERE SO FEW INSECTS IN THE SEA?

Our planet is the watery one—the only blue and white planet in our solar system. More than three quarters of the Earth's surface is covered by either water or ice. Seen from afar, large swirls of water droplets form white clouds sweeping across our blue planet. There really is a lot of water on this planet of ours, and that's where life began: within the sea. Walk along a seashore and take a good look at the diversity of what's washed up along the beach. Don't look at the numbers of species; instead, look for the number of phyla. Phyla are the really big divisions of the animal world; they range from fishes, crabs, starfish, jellyfish, and seashells to a goodly number of worm-like creatures. All the animal phyla have representatives living in the sea. But where are the insects?

Insects belong to a large phylum of animals called the arthropods. Having jointed legs, they include the crabs, lobsters, horseshoe crabs, spiders, scorpions, centipedes, and millipedes—and the insects. Many arthropods live in the sea, but only a very few of them are insects. There are a few spiders, but no millipedes or scorpions are marine. Water beetles and mosquito larvae spend a good part of their lives under water, but they do this in fresh water. Insects live almost entirely on the land or in land-based aquatic ecosystems—why is this so?

Insects are a terrestrial lineage, and if they did evolve in water it was in fresh water. Based on the most primitive insects living today, it appears that they originated in ponds and streams—not from along the ocean shore. The richness and complexity of a land biota allowed insects to diversify more abundantly than any other living lineage. A simple reason for insects not having ventured into the sea is the vast array of predators waiting beneath the waves.

In similar fashion, vertebrates pioneered the land from streams and estuaries, not at the ocean shore. Further along in the history of life, mammals originated from terrestrial mammal-like reptiles, and since that early time only a few mammalian lineages have become aquatic. Among these are otters, beavers, and hippos in freshwater, and whales, dolphins, seals, and sea lions in marine habitats. Regardless of how much time these mammals spend in the water, all need to fill their lungs with air from above the water's surface. Unable to breath underwater, they betray their terrestrial origin. Next let's leave the Animal Kingdom and look at another lineage of terrestrial life-forms with high species numbers—the flowering plants (what botanists call the Angiosperms).

ANOTHER BIG CROWD: THE FLOWERING PLANTS

Angiosperms—flowering plants to most people—are not only numerous in species, they come in a grand variety of sizes. In the case of most beetle species, you can stuff a thousand of them in a gallon jug and still have space left over. Most flowering plants, in contrast, come in larger sizes.

The smallest flowering plants are the floating freshwater duckweeds (Lemnaceae family), ranging from about the size of an O in this particular font to an inch or two in length. With only about thirty species, they are the bottom end of the size scale in flowering plants. At the other extreme, Angiosperms include eucalyptus trees up to three hundred feet (97 m) in height, baobab trees with trunks as wide as a truck, and the double coconut palm with leaves up to thirty-five feet (11 m) long. Simply stated, *flowering plants are the most diverse group of larger organisms that ever lived on planet Earth!* Whether small grasses or tall trees, flowering plants are major players in terrestrial ecosystems. Fish also come in a wide variety of sizes, with whale sharks reaching forty tons, but fish species number around 30,000 species. Mammals include the largest creature ever—the blue whale—reaching one hundred feet (30 m) in length and weighing over one hundred tons. But living mammals number around 5,500 species.[13] In contrast, recent estimates place the number of described flowering plants at around 282,000 species. And with around a thousand new flowering plant species described every year, they are bound to get to 300,000.[14]

Clearly, when you combine size and species numbers with the food production of photosynthesis, the flowering plants are the biggest game in town! They include the grass family (10,550 species), the legume family (19,500 species), the exquisite orchids (22,500 species), and the sunflower family (23,600 species).[15] But keep in mind, high species numbers may not add up to a major environmental presence. Orchids tend to have relatively few individuals in most plant communities. When discussing biomass, orchids are an insignificant part of natural landscapes—even in cloud forests where many sit as epiphytes on trunks and branches. Grasses are different, often dominating broad landscapes with billions of individuals. Many parts of the world are called grasslands for this very reason, including prairies, steppes, savannas, and mountain-top alpine communities.

With their generous numbers, their varied sizes, and a complex architecture, flowering plants are the primary determinants of vegetation structure in a majority of today's ecosystems. Not only are they significant for their three-dimensional forms, but also as sources of nutrition. Since all but a few flowering plants are green and photosynthetic, they stand at the base of local food chains. It is these—and other green organisms—that convert solar energy into nourishment for all of us. In fact, as flowering plants expanded their numbers over the last 100 million years, other lineages have been able to expand their numbers as well. A recent study of beetles clearly showed that those beetle lineages feeding on flowering plants have achieved greater numbers than beetle lineages using other food sources.[16]

Conifers are not flowering plants, and these trees dominate many colder and seasonally stressful habitats. However, while pines, firs, redwoods, junipers, podocarps, araucarias, and other conifers can have very high numbers of individuals in a great variety of forests and woodlands, they total only around one thousand species worldwide. And, though they dominate cooler northern and many montane forests, they are all trees; only their juveniles are important in the understory. More significantly, conifers defend themselves with a strong chemical arsenal. That's why conifer wood is such great building material, why very few animals feed on pine needles, and why the understory of conifer forests is so poor in species. There's simply not as much to eat in a conifer forest. In contrast, flowering plants invest less energy in self-protection and more in growth. In a sense, they "live fast and die young," powering a more nutritious vegetation. Beyond flowering plants and conifers, ferns are another important and species-rich lineage of land plants. Ferns and the fern allies (such as lycopods and horsetails) number around 12,000 species. Though only a very few are tree-like, many species are common in moist forest understories, and some are an important component of the epiphytic community in moist forests. Because they reproduce with swimming sperm, ferns, mosses, and their allies are restricted to evergreen or seasonally moist vegetation. All told, flowering plants are both the most numerous of land plants, the most nutritious, and the most colorful.[17] We will return to the subject of land vegetation and species numbers in chapter 4 and discuss the role of the flowering plants in making terrestrial ecosystems ever more complex in chapters 7 and 8.

As we've noted, plants and animals have many more species in terrestrial environments than in the oceans. But this generalization may not hold true for that kingdom of living things, which includes the smallest forms of life: the bacteria. No matter whether we are counting the numbers of individuals, numbers of lineages, variety of biochemical pathways, or varied abilities to survive in extreme environments, the bacteria rank first on planet Earth. Next, let's take a closer look at life's most successful crowd: the bacteria.

Chapter 2

BACTERIA, EUKARYOTIC CELLS, AND SEX

I f we want to talk about really huge numbers of living organisms, we will have to discuss the **bacteria** (singular: bacterium). These are the world's simplest living things, and the most numerous. They are multitudinous in the habitats that we are familiar with, but they also survive on the ice fields of Antarctica and in rocky fissures deep within the Earth. Bacterial activity supports strange sightless animals living at fumaroles on the bottom of the sea, and bacteria are an important component of the phytoplankton that captures the energy of sunlight near the ocean surface. Trillions of bacteria, of several hundred kinds, live within our intestinal tract, helping us digest our food, providing us with essential vitamins, and helping us defend ourselves against disease (though the wrong kinds can kill us).

Recently, using newly developed DNA screening, microbiologists have also discovered that there are huge numbers of bacteria in the sea and in the soil. Earlier, we would culture bacteria on petri plates to find and identify them, but many bacteria cannot survive on a dish of agar, remaining undetected and unknown. Using recently developed DNA surveys, we are now able to identify bacteria without having to culture them. One study claims that there may be as many as 10^{10} (10,000,000,000) bacteria in a gram of dirt.[1]

In discussing bacteria, we are being quite specific. Bacteria may be microbes, but microbes include a great variety of other minute organisms as well. Protozoa and flagellate amoeba are microbes, but these have larger cells containing nuclei; they are fundamentally different from bacteria. With few exceptions, bacteria are so small that we cannot see them without the aid of a microscope. Before going further with bacteria, however, let's dispense with the even smaller "life form" called viruses.

Viruses do not truly qualify as living beings. To make sense of that statement, let's think about the operations that define life. First of all, living things have to have some way of keeping themselves intact in a changing environment. Bacteria are encased in a tough wall to protect themselves, and many can even transform themselves into tough spore-like devices when the going gets really rough. Just as important is the ability to capture energy in order to carry on life activities. Food, digestive enzymes, and metabolic systems help keep the bacterium supplied with energy. Another key responsibility for all living things is that they must have a way of replicating themselves. This necessitates a genetic program that can be duplicated and transferred from generation to generation. The long double-stranded DNA helix, held together by precise base pairing, is the principle information storage and duplication mechanism for life on Earth. By having the double-stranded DNA helix unwind, each separate strand can then be duplicated to form two new double-stranded helices. These then become the hereditary information for two newly divided "daughter cells" in the process of reproduction.

Viruses do carry short strands of DNA (or RNA) for replication, but they lack two major life criteria. They cannot garner energy for themselves, nor can they reproduce themselves. Then how can they make us so ill? Viruses seem to be a horrible example of "**Murphy's Law**," where something that *could go wrong did go wrong!* Viruses appear to be rogue snippets of DNA or RNA—the basic stuff of genetic information—that got themselves a proper protein coat "and learned how to travel." What these snippets of DNA or RNA do is to travel around until they encounter a cell they can parasitize, attach to that cell, and insert their short genetic program into the cell. Once within the cell, the viral genes commandeer the gene-replication machinery of the infected cell, which then begins to churn out more viruses. What went wrong here—in the sense of Murphy's Law—was that a small fragment of DNA or RNA got itself a proper encasement, allowing the fragment to travel, to attach, and to infect specific host cells. Viruses are the simplest of parasites; they are "alive" only in their hosts, and each kind of virus is limited to a narrow range of cells it can infect. Among humans, viruses are responsible for influenza, polio, the once-devastating smallpox, and AIDS (caused by the HIV virus). Viruses are so simple that they can even be crystalized and their structure studied by X-ray diffraction. Parasitism, moreover, is a lifestyle that many other living things have also adopted, including bacteria, protozoa, many fungi, a number of animals, and even a few plant species. In many of these, drastic reduction in body size and loss of complexity has occurred, as the parasite depends more and more upon the host. However, and unlike viruses, these larger parasites reproduce—often prodigiously—on their own. But enough of parasites, let's get back to the world's smallest form of independent life: bacteria.

MICROSCOPIC BUT SUCCESSFUL

The most singular aspect of bacteria is their very small size and the lack of organelles within their cells. There are larger bacteria, especially in the lineages that carry on photosynthesis, but most bacterial cells are about one-tenth of the length of an average animal cell. (Bacteria average 1–2 microns thick and 1–4 microns long. A micron is a millionth of a meter, or a thousandth of a millimeter. There are 25.4 millimeters in an inch.) Considering that many bacterial cells are similar in all three dimensions, and that they are about one-tenth the size of animals cells, we can multiply one-tenth (length) by one-tenth (width) by one-tenth (depth), with the result that most bacterial cells have only one-thousandth of the volume of an average animal cell. This is a difference that accounts both for the ubiquity of bacteria and for limitations in what they have been able to accomplish.

Despite their small volume, different lineages of bacteria display a wide array of biochemical activities—more than in all the rest of the living world. Surely this is due to the fact that they have been active here on Earth for a longer period of time than any other lineage—a span of over three billion years! To gain nourishment, most bacteria exude enzymes that digest the food around them. They then absorb their digested dinner. Not a very efficient way to get chow, but it works well in a wide variety of environments. Just as important, each bacterial cell carries a genetic code of DNA, allowing it not only to repair and maintain itself but also to replicate that code for future generations. Efficient energy harvesting and effective reproduction are essential to all forms of life.

Studies of ancient rocks suggest that some kind of bacterial life was in place on planet Earth 3,500 million years ago. These bacteria must have gained energy from a variety of chemical reactions. Breaking down sulfur compounds was an early strategy, and such sulfur-utilizing bacteria are still with us today. But finding just the right energy-rich substrates limited the expansion of early bacterial life. A grand advance took place when one lineage of bacteria developed the biochemical means to capture the energy of sunlight. We call this process **photosynthesis**: transforming the energy of sunlight into the energy of chemical bonds. The world's earliest photosynthetic bacteria used compounds of hydrogen and sulfur to gain the hydrogen atoms they needed to construct carbohydrates. Transforming the energy of sunlight into chemical energy requires a complex system, including chlorophyll molecules and additional pigments, all within an elaborate molecular framework. By using sunlight to build carbohydrates and other complex molecules, photosynthesizing bacteria became food for others.

Early photosynthesis was a major advance in the history of life, but finding appropriate sulfur molecules limited this process. Then came a world-changing advance: water-splitting photosynthesis! This was accomplished by the **Cyanobacteria** (called blue-green algae in older literature). Cyanobacteria can be much larger than average bacteria and are often connected in long filaments big enough to produce the dark green glop and yellow-green mats we often see in ponds and streams. At the other end of the size scale sits *Prochloroccus*, less than a micron in size. So minute, this marine cyanobacterium was not described until

1988! Its photosynthetic ability is essential to life in the ocean, its minute size probably a response to filter-feeding predators.

What the Cyanobacteria accomplished was phenomenal: they were able to *rip apart the water molecule* in order to get the hydrogen needed to build carbohydrates. This new kind of photosynthesis utilized the hydrogen of water, expelling free oxygen into the air. **Oxygenic photosynthesis** was probably operational by 2,700 million years ago.[2] Coupling the hydrogen of water with carbon dioxide by using the energy of sunshine, Cyanobacteria empowered a vastly larger biosphere and—with time—gave rise to breathable air!

Another biochemical triumph, **nitrogen fixation**, is also exclusive to the bacteria. This resembles photosynthesis in an important respect. Pulling the water molecule apart in photosynthesis is no ordinary reaction. Hydrogen and oxygen really like each other. When they get together there is a hot time, as the burning of the hydrogen-filled Hindenburg zeppelin displayed. But once together, the two hydrogen atoms are tightly embraced by oxygen, and they want to stay that way. Pulling them apart takes a lot of effort. The biochemical process of nitrogen fixation faces a similar challenge. Our atmosphere is full of nitrogen, to the tune of about 79 percent nitrogen. So what's the problem?

The problem is that the nitrogen in air is made up of *molecular nitrogen.* Here, two nitrogen atoms hold each other close with three covalent bonds, and they like to stay that way. Because of this strong bonding, molecular nitrogen remains inert to most life activities. And this is why "nitrogen-fixing" bacteria are so important: they can separate the two nitrogen atoms and build them into nitrates or ammonia. Once in this form, other living things can use the "fixed," or "available," nitrogen for their life activities. Because nitrogen is a central component of nucleotides (in DNA and RNA), amino acids (in proteins), alkaloids, and more, all living things require accessible nitrogen for growth and reproduction.

HOW DO BACTERIA MAKE MORE BACTERIA?

Bacteria multiply by division. That is, the individual cell divides into two, forming similar daughter cells. Before this important aspect of the life cycle takes place, the bacterium has to have been well-fed and in an agreeable environment. During this time of growth and prosperity, the thin circular chromosome within the bacterium will have made a duplicate copy of itself. (It is the **chromosome** that carries the long double helix of DNA, the genetic information for keeping them running). The two thin filament-like chromosome loops will then attach themselves at the same point on the inside wall of the bacterium. This single attachment point will split, and here is where the bacterium will pinch in two. Some bacteria carry two chromosomes but divide in much the same way. All told, both daughter cells will come away with a complete chromosome complement, carrying all the genes needed to continue being a busy bacterium. In fact, when conditions are good, many bacteria can grow and divide every twenty to thirty minutes. With plentiful resources, bacteria can multiply exponentially, as rotting food often makes clear.

But did you notice something else? We've been discussing bacterial reproduction and we haven't mentioned sex. That's because there isn't any. There are no male bacteria searching for female counterparts in this story. There is only the simple internal duplication of the chromosome into two, followed by a separation of the two strands to become the chromosomes for each of the two new daughter cells. Simply stated, bacteria do not have sexual reproduction as more complex creatures do. Yet we all know that bacteria are highly adaptive—they mutate frequently, and new strains can make us very ill. How can they be so adaptable without having their genetic material scrambled around in the process higher organisms use?

Turns out, bacteria do have ways of getting together, called conjugation. In this closely positioned state, they can exchange bits of genetic information (DNA) and that's about as close they get to having sex. They can also exchange or acquire packets of genes from other bacteria via plasmids, or by virus-like intermediates. In other words, bacteria do have effective mechanisms for exchanging sections of DNA, thus gaining new hereditary information, new variability, and new possibilities. All this is in addition to naturally occurring mutations as they grow and divide. Though the vast majority of mutations may be deleterious, a very few will give an advantage, helping in the struggle for survival. Clearly, we must monitor bacterial diseases constantly; our foe is ever changing.

Despite extraordinary biochemical diversity, fast life cycles, thousands of different lineages and diverse genetic strains, the bacteria remain sharply constrained. Because they are so small, and because they carry only one or two chromosomes, they are limited in the amount of information they can carry. If they pick up packets of new genes from other bacteria, they cannot just keep making their chromosome longer. They must, in fact, jettison some genetic information when they get too much. They're simply too small to carry around an expanding file cabinet. Remember that these things are really small; the period at the end of this sentence provides enough area for about ten thousand average-sized bacteria to congregate.[3] Our skin hosts an estimated three hundred million bacteria, and our large intestine is thought to be teeming with around seventy *trillion* of them.[4]

Because they are so small and so versatile, bacteria are impossible to stay clear of. True, some of these bacteria can cause disease or reduce the efficiency of animals, which is why our hog farmers and cattle ranchers add tons of antibiotics to animal feed. But, over the longer run, this may be stupid! Yes, these antibiotics improve the growth of the animals we eat and, in that way, make meat more affordable. But there are two serious reasons that these short-term strategies might backfire. The first is simple: by using large quantities of antibiotics in raising our livestock, we are creating environments saturated with antibacterial compounds. And you get only one guess as to how Mother Nature is going to respond: natural selection! This is a gargantuan program screening for resistance to antibacterial medications, the same agents we use in guarding ourselves from bacterial attack. And then there is a second problem. We are learning that low levels of bacteria may help protect animals from microbial attack! Thanks to intense study of the laboratory fruit fly (*Drosophila melanogaster*), we have discovered that a persistent low-level bacterial commensal within the fruit fly protects the fly from further serious microbial infection! Since insects have been around for over three hundred million years—subject to bacterial attack over that entire length of time—it makes sense that they have developed a lot of defenses to deal with microbial challenges. Extrapolating from insects to other animals and ourselves, investigators have found that at least some "native microbes are symbionts that shape our immune response and help us stay healthy."[5]

We've mentioned how bacteria can transfer DNA between themselves. Interestingly, it turns out that many bacteria can pick up packets of genes from bacteria to which they are not closely related. This is called *horizontal gene transfer*, a phrasing that reflects the metaphor of an "evolutionary tree." Such a tree places contemporary living species at the ends of distal twigs, with their earlier ancestors along the branches, and their deep evolutionary origin along the trunk. On such an evolutionary tree, with an ancient trunk and more modern branches, contemporary gene transfer between unrelated species appears to be "horizontal" across distal branches. In contrast, complex animals get their genes from parents, who received their genes from their parents in an "up-from-below" pattern—not horizontally from our contemporaries.

Effective horizontal gene transfer is rare in more complex living things. Rose breeders can only use other roses for their breeding program. Until the advent of genetic engineering, there was no way of getting a petunia gene into a rose. The same was true for animals; animal breeders were stuck with the close relatives of the animals they were breeding. And that's why antelopes all look like antelopes; they are highly restricted in their choice of mates. Because horizontal gene transfer is so rare in higher organisms, we can talk comfortably about thousands of beetle species. Lacking horizontal gene transfer, each beetle keeps looking like all the other members of his or her species.

But in the case of bacterial species, how can we distinguish species among bacteria when they are busy tossing genes back and forth between lineages? With so many bacteria in so many places, with such a long history, and with so many looking so similar, how can we recognize, designate, or even hope to find species in the world of bacteria? This question brings us back to the fundamental unit of biological diversity: the species.

DO THE BACTERIA REALLY HAVE SPECIES?

The **biological species concept** in higher plants and animals goes something like this: "A species is a population (or series of populations) of interbreeding individuals that cannot exchange genes with other such species."[6] In other words, a species should be *genetically isolated* from other closely allied plants or animals. Whether the isolation is by long distances, inability to mate, or offspring that are sterile, the result should be the same: no gene exchange with individuals outside the species! In an ideal world, a species is a collection of plants or animals that are advancing through time, down the evolutionary road, all by themselves. Genes from outside sources simply can't get into this species, to mess up its morphology, or obscure its past. Philosophically, the biological species concept is a great idea, and it works well for a large percentage of plants and animals. Unfortunately, the biological species concept seems inappropriate for the world of bacteria. They can and they do exchange genes across different lineages. Our common intestinal bacterium (*Escheria coli*) is thought to have some 25 percent of its genome acquired from other species.[7]

The sad truth seems to be that we cannot use the generally accepted concept of *species* when we are talking about bacteria. If you can pick up genes from far and wide, you have extraordinary potential for doing new and interesting things. But after you have acquired those distant genes, you may suddenly become quite different from even your own ancestors. Bacterial species don't seem to have a tree of close relationships. Rather, they seem to be moving through time as if they were on an interconnected trellis.[8] However, bacteria are important; they are a huge component of the living world. If we want to understand them, we must have a means of classifying them. And though the horizontal-gene-transport problem obscures some of their relationships, perhaps we can still find effective ways of arranging the world of bacteria.

When it was discovered that some human diseases were caused by bacteria, the medical community had to find ways of identifying and classifying these critters. In a medical setting, getting a bacterial identification correct could mean the difference between life and death. Some bacteria do differ clearly in form or color, such as the spiral Spirochetes, the small spherical Cocci, or the larger filamentous Cyanobacteria. These were easy to segregate. But the vast majority of bacteria are little rod-shaped things, differing little in appearance. An important test was to see how the bacterium responds to specific staining reagents. Gram-positive bacteria were distinguished from Gram-negative bacteria by how their walls reacted to stains. Also, biochemical and nutritional tests were devised as means of identification. Growing the unknown bacterium on agar with different nutritional qualities was a way to tease out their identity. This may not produce a classification of clear "evolutionary branches," but it worked well enough to help identify them. Now, with piles of new DNA data, we can seek those genetic traits that are basic to the working of the bacterium and unlikely to be easily exchanged. We may not get clear-cut species out of this activity, but we will develop a better classification.[9] More important, these studies revealed a big surprise: the so-called "bacteria" included two very different assemblages!

A major breakthrough in the understanding of bacteria came about after Carl Woese and his colleagues at the University of Illinois found that bacteria really comprised two profoundly different groupings. These groupings were first separated as the Eubacteria and the Archaebacteria. Further studies confirmed this dichotomy, and the two are now called the **Bacteria** and the **Archaea**. Certain gene complexes and biochemical traits made clear that Archaea weren't paddling the same boats as were the other bacteria. For one thing, the cell walls of Archaea differ in chemical structure from the walls of other bacteria. In addition, RNA transcription in Archaea differs significantly from transcription within the Bacteria. Even more striking was the fact that many of the Archaea are found living in extreme environments. These ranged from the steaming water of hot springs to the ice fields of Antarctica, and from sulfur-belching fumaroles on the deep ocean floor to water of extremely high salinity. Perhaps it was their ability to live at very high temperatures that made scientists assume these microbes might be ancient survivors of Earth's earliest history, when our oceans were near the boiling point. Nevertheless, and regardless of their origins, the more the Archaea were studied, the more obvious it became that they are distinctive. None of the Archaea are human pathogens and, as we've just noted, many are *extremophiles*. Today, they are seen as one of the three major living domains or "super kingdoms" of life: Bacteria, Archaea, and the Eukaryota.[10]

ARE THE ARCHAEA REALLY ARCHAIC?

As more data was gathered regarding the three large domains of life, odd similarities were noted. Clearly, eukaryotic (nucleated) cells had some genetic traits that were closely similar to those of the Bacteria. One of the grand surprises at the end of the late twentieth century was that we could put a human gene into a bacterium and have that bacterium *produce a human enzyme!* Today, we are busy manufacturing human insulin for diabetics in exactly this way! No one had expected this. After all, our lineage and the bacterial lineage have been separated for at least two billion years. Apparently, Mother Nature has been following the old maxim: "If it works, don't fix it!" Our ability to get bacteria to make human insulin makes clear that some of our basic biochemical-genetic machinery has remained unchanged over billions of years. These biochemical facts are strong evidence for evolutionary continuity since early bacterial times. However, as regards RNA transcription and some other features, the Eukaryotes resemble the Archaea more than they do the Bacteria. It appears that Bacteria and Archaea have both played a role in the later origin of eukaryotic cells.

Let's back up a minute here and review our nomenclature. The cells of **Eukaryotes** have a nucleus; their name declares "true nucleus." **Prokaryotes** have no nuclei, and their name means "before the nucleus." The prokaryotes include all the Bacteria and the Archaea. In addition to not having a nucleus, prokaryotes do not have organelles within their little cells. The word prokaryote implies that these cells existed *before* the more complex eukaryotic cells evolved. Indeed, the fossil record supports such an inference (more about this in chapter 8). Not only are bacteria cells generally much smaller than eukaryotic cells, they are not as complex. Though they carry on a huge variety of different biochemical activities in their many different lineages, and are found in every nook and cranny on the planet, the individual bacterium is quite limited in what it can do.

Fossil evidence for the earliest forms of life is meager, and minute eukaryotes are difficult to distinguish from bacterial-grade life. It seems likely that the "eukaryotic cell" may have emerged around 1.5 billion years ago, a full two billion years after bacterial life began. Whatever the actual dates, the greater complexity of eukaryotic cells was a later development in the history of life. But what about the two grand divisions of the prokaryotic world? Those bacteria that survive in the hottest water—close to the boiling point—are nearly all Archaea. While some Bacteria can live in water up to about 180°F (82°C), few can handle the turbulent heat in which some Archaea thrive. Keep in mind that temperature is a measure of molecular motion. The higher the temperature the greater the motion, which is why complex organic molecules begin breaking apart above the boiling point of water and why metals melt in thousand-degree heat. Staying together and carrying on one's life activities at the boiling point of water is no simple task. Nevertheless, some scientists thought that life might have *originated* under such fierce conditions, and that the Archaea were the living descendants of those earliest forms.

Not so, claimed Thomas Cavalier-Smith of Oxford University. In bold contrast to the received wisdom of earlier work, Cavalier-Smith argues that the Archaea are, in fact, a more modern assemblage of bacteria.[11] Because the Archaea have RNA transcription protocols and some other characters found in the later-evolved Eukaryota, it would seem logical that the Archaea are also a more modern development. More telling may be the fact that some Archaea live at temperatures that no other living things can tolerate. Does it seem reasonable that the earliest life forms developed under such severe conditions? Just as polar bears and penguins are recently evolved residents of our most severe polar climates, it would also seem that Archaea were a later development in the history of life. Despite their name, the Archaea may not be especially archaic. Furthermore, Professor Cavalier-Smith addressed an even more fundamental question: Why haven't the Bacteria been more progressive?

Why have bacteria spent more than three billion years being so small? True, they're hugely successful in biochemical diversity, as well as invading every corner of the globe, but why haven't they been able to do anything along the lines of plants and animals, or even protozoa? Bacteria do have an ability, called *quorum sensing*, that lets them know about others of their kind nearby. And they can form some larger aggregations in this way. Also, Cyanobacteria can form long filaments made up of very large cells. But this is nothing comparable to a simple plant or animal that is, in fact, a functional union of many millions of differing eukaryotic cells. What is it that has kept the bacteria from becoming more than just bacteria?

Professor Cavalier-Smith claims that one of their most important traits has kept bacteria trapped in an evolutionary vise from which they have been unable to escape. That vise, he argues, is the bacterial cell wall itself. This wall must protect the interior of the bacterium from toxic substances or high concentrations of chemicals in the exterior environment. It must prevent the leakage of essential substances from within the bacterium. It must sense the environment and keep the interior of the cell informed of danger. The wall must also allow for the secretion of digestive enzymes and waste products, even as it enables the absorption of nutrients. All this is in addition to being able to pinch itself in two during division, and repairing itself in the face of environmental stress. The bacterial wall allows the bacterium to remain a sensitive and dynamic living being in a huge variety of inhospitable environments. Obviously, when you are enclosed within something as solid and efficacious as the bacterial wall, any change in that protective armor might prove disastrous. This, Cavalier-Smith argues, is why the bacteria have remained so consistent: their protective encasement has not permitted them to expand and enlarge.

The importance of Cavalier-Smith's argument is that it helps us understand why, over more than three billion years, the bacteria were unable to do what the larger, more complex eukaryotic cells have been able to do—become the building blocks for larger and more complex life forms. Unable to change the nature of their walls, a defensive bulwark against an often hostile world, bacteria are doing today most of the same things they've been doing for a very long time.

Bacterial success combined with bacterial stasis reminds us that there are two general ways of increasing biodiversity. The first, as in the bacteria, is to achieve a certain level of complexity and then diversify "laterally," filling more and more niches with additional variants of your kind. The second, and a major theme in our story, is to increase one's inherent complexity and diversify from this new state. A larger, more complex cell, in fact, proved to be the platform required for building increased biological complexity on planet Earth.

A MAJOR ADVANCE: THE EUKARYOTIC CELL

A revolution in the nature of the cell wall, in fact, ushered in a major new chapter in the history of life. This new kind of cell wall allowed the ancestors of eukaryotes to *engulf* their prey, just as protozoa do. Swallowing one's dinner is a lot more efficient than exuding enzymes and then absorbing the digested food through the cell wall. A more flexible and more versatile cell wall did more than help early eukaryotes swallow dinner; this was the beginning of entirely new possibilities. Surely the most profound innovation was forming an intimate and permanent relationship with an energy-processing bacterium. By engulfing a new bacterial partner that became the **mitochondrion**, the eukaryotic cell was empowered to achieve new levels of complexity. Because the mitochondrion is about the size of a bacterium, and has interior membranes that resemble bacterial membranes, early biologists speculated that the mitochondrion had once been an independent bacterium. Lynn Margulis championed this idea in the late 1960s, calling it **endosymbiosis.** She suggested that a close symbiotic partner—the bacterium—finally became a necessary organelle within the much larger eukaryotic cell itself. The term symbiosis implies that both partners are benefiting from the relationship, though it could be that the bacterium was essentially enslaved by the larger eukaryotic cell.[12] Either way, slave or partner, mitochondria became a central feature of the eukaryotic cell. Margulis's endosymbiotic hypothesis found confirmation with the discovery that mitochondria still carry a few of their own genes—evidence that they had once been independent life forms. By providing the eukaryotic cell with a more efficient metabolism, the acquisition of the mitochondrion was an essential early innovation in the advance of complex life forms.[13]

A new, more flexible wall structure may have been the first fundamental step forward, but getting more efficient energy processing from the mitochondria opened the door to other advances. Think about the small bacterial chromosome. When bacteria acquire new genetic information, they often have to jettison part of the information they're already carrying; they simply can't handle an additional load. But with a larger cell, and more efficient energy delivery, the eukaryotic cell was able to build, maintain, and duplicate much larger file cabinets! This cell could carry not one chromosome but lots of chromosomes. Neatly enclosed within the nucleus of the larger eukaryotic cell, chromosomes and their genetic information were protected from the rough-and-tumble dynamics in the cytoplasm (the aqueous cell contents outside the nucleus). Cellular dynamics include building and destroying proteins as needed, renewing and repairing cell constituents such as enzymes that regulate reactions, digestion of food particles, and preparing for cell division. Unlike bacteria, eukaryotic cells have "motor proteins," which move materials around within the larger cell. All this takes energy, pro-

advance in the escalation of biodiversity on planet Earth. In addition, red algae, brown algae, and a few other lineages were the benefactors of yet additional symbiotic unions within their cells.[16] However, because of stiff cellulose cell walls, plants have been limited in their ability to form varied morphological forms, unlike animals with their thin flexible cell membranes. But whether plant or animal, the more complex eukaryotic cell has a problem: how to divide and reproduce.

HOW EUKARYOTIC CELLS MULTIPLY

Before moving to the world of eukaryotic creatures, we must examine another fundamental aspect of the eukaryotic cell: replication. We need to understand how the eukaryotic cell divides and multiplies—a process called **mitosis**. The first sign that a eukaryotic cell is going to divide is the disappearance of the nucleus, as the nuclear membrane is dissolved. At the same time, the chromosomes condense into thick visible structures. During the regular activities of the cell, chromosomes are not visible under a microscope. Even staining does not show them. But as cell division begins, the chromosomes condense, becoming dark little sausages under the microscope. (We observe this process after the cell has been killed and stained.) The chromosomes move toward the center of the cell and align themselves along a central plane within the cell (called the metaphase plate). This is the same plane along which the cell will divide, or pinch itself in two. So now we've got all the chromosomes hanging out along this saucer-like area at the center of the cell. They then begin to align themselves in such a way that each chromosome will separate into two strands, with the two strands ready to move in opposing directions. Lines appear to form within the cell, reaching toward the center of each chromosome. Like stars, these lines radiate inward from two distal poles, and it is along these lines that the sister strands are pulled in opposite directions, and the cell is ready to divide down the same plane on which the chromosomes had aligned themselves. Voila—two new cells, each with the same number of chromosomes as the original cell!

Before cell division could occur, preparations were made in advance. During the period of time when the cell was going about its normal business, it was also doubling its chromosome strands. Each double-helix strand of DNA had to be transformed into two double-helix strands. Thus, when cell division began, the chromosomes were already two-stranded. Let's review this process again: Once mitosis begins, the nucleus disassembles its membrane, freeing the chromosomes to move to the center of the cell and align themselves properly. Set to travel in opposing directions, each of the two sister DNA strands separate and head to one of the two poles of the now-dividing cell. At the same time, mitochondria and other cell organelles are also dividing, to ensure that each daughter cell has what's required to be a successful new cell. In short, mitosis produces two identical cells where there had been only one. And, with over a trillion cells making up a human being, you can understand how critical cell division is to our growth and health. Since many of our cells are being continuously lost, whether on our skin, in our bloodstream, or along the walls of our intestine, cell division maintains our active bodies even as we age. But this is just one aspect of the game of life. Eukaryotic organisms have another game they like to play—we call it sex.

WHAT GOOD IS MEIOSIS AND SEX?

Meiosis is the starting gun of sex. Meiosis is the process of cell division that produces sex cells (gametes) in eukaryotic organisms. Meiosis differs significantly from mitosis, the common form of cell division we've just been talking about. Looking at stained cells, killed during the process of meiosis, one sees much the same activities as in mitosis. However, there are two main differences. The first is that meiosis is a two-stage process, comprising two cell divisions. The second difference is how the chromosomes line up and then divide during their first cell division. In meiosis, before lining up on the metaphase plate, *homologous chromosomes form pairs*. Most organisms received one set of chromosomes from their female parent and a second set from their male parent; thus their cells contain two sets of chromosomes (the **diploid** condition). Humans carry forty-six chromosomes in their cells; twenty-three came from Mom, and twenty-three came from Dad. Each of these twenty-three chromosomes is different from the other twenty-two, and each will link up with its own homologue during meiosis. Thus, instead of having forty-six chromosomes crowding the center of the cell, in meiosis we've got twenty-three pairs ready to divide and head in opposite directions. Thus, the first division of meiosis gives rise to two daughter cells with *half the normal adult number of chromosomes:* twenty-three in the case of humans (the **haploid** state). The second division of meiosis follows in the manner of mitosis and results in four cells, all with half the regular chromosome number. In animals, meiosis usually produces four sperm cells, but in the female line meiosis often results in one very plump egg cell and three skinny sisters that are cast aside. Unfair but pragmatic: the larger female egg cell will have more nutrition to begin a new life. The smaller male sperm carries only enough nutrition to power its flagella, swim to the egg cell, and donate its haploid genome for the new diploid organism.

Meiosis takes ordinary cells with two chromosome sets—the diploid condition—and reduces them down to sex cells with only one set of chromosomes: the haploid condition. Meiosis yields sex cells that are utterly and completely useless (in most organisms) until they get together with another sex cell to restore the diploid condition. Consequently, most eukaryotic cells in larger organisms are **diploid:** they have two chromosome sets within the nucleus of each cell. There are exceptions (this is biology): many algae, mosses, and liverworts live most of their lives in the haploid condition. Nevertheless, sperm and unfertilized egg cells represent the haploid part of the life cycle in most eukaryotes. (We'll discuss the chromosome compliment of a plant or animal—**ploidy**—in the next chapter).

WHY IS SEXUAL REPRODUCTION ADVANTAGEOUS?

Why produce sex cells; cells that are useless until and unless they unite with another haploid sex cell? This has been a central conundrum for biology over several centuries. Why waste time and energy making sex cells when complex organisms could simply divide in two and produce little copies of themselves? Why should a huge Alaska brown bear mother do all the work of bearing young, nursing them, and teaching them, while brown bear males do little more than defend their territories? Like most animals, date palms also come in two sexes. Again, it is the female palm that produces all the sweet dates we are so found of; male date palms produce nothing more than large inflorescences and a lot of pollen. The troublesome question is: *Why is sex so widespread in the living world?*

Biological systems are complex systems—the products of many processes and many interactions, all with a very long history. Again, this isn't physics. Two fundamental explanations for the existence of sex are found in the process of meiosis itself. Recall that homologous chromosomes must pair together before becoming aligned on the metaphase plate. Precise pairing of strands in meiosis is necessary for the accurate separation into the two diverging daughter cells. Just as significantly, this pairing allows for *chromosome repair* in a way that ordinary mitosis does not.

The second advantage of precise pairing is called *crossing over*. This allows sections along the paired chromosomes to be exchanged—a terrific way to get new gene combinations onto the same chromosome. Imagine a chromosome with a bad gene at point **q**, while its chromosome partner has a bad gene at point **t**. With crossing over there is the possibility of forming a single chromosome with good genes at both points **q** and **t**. Of course, this results in the other chromosome getting stuck with both bad genes. Tough luck! The good news is that we've now got a new combination, with both good traits on the same chromosome. What meiosis does is allow both for repair and recombination. And that's just the beginning.

Sexual union, when sperm and egg unite, brings together two gametes from two separate individuals, who may be quite different themselves. Each of the parent's sex cells (gametes) is a sample of that parent's genetic resources. Think of this in terms of your grandparents. Your parents gave you a *sampling* of your four grandparent's chromosome sets, via their sperm and egg. Theoretically, your parents could have had two million children, each with a different sampling of their four grandparent's chromosomes. No wonder families can have such markedly different children!

Now let's expand our view to look at sex from the perspective of the entire species. Sex means that the genetic heritage of a species can be reshuffled every generation, rather like a huge deck of playing cards. From the perspective of a large population of individuals, sex constantly produces individuals with both old and new gene combinations. This has several profound advantages. Genetic diversity means that a species can have individuals doing well in both drier and wetter, or warmer and colder, parts of its habitat. Constant genetic reshuffling also means that a species is more likely to adapt when there is a change in climate. And finally, a diverse and constantly variable species may be the only way to withstand incessant attack by parasites and pathogens. It is not "Lions and tigers and bears, oh my!" that are likely to end your days. Rather, it is these smaller, quickly changing enemies, which cause devastating losses among plants, animals, and people. A sexually reproducing host species is more likely to contain individuals who can survive in the face of attack by a new pathogen. Sex, and the constant scrambling of genetic combinations, may be the only way for a species to survive in a hostile world of unpredictable climate perturbation and ever-variable pathogens.

Fundamentally, sex serves the population and species—not the individual! Meiotic division and sexual union are constantly shuffling the cards to produce a grand variety of hereditary outcomes. Those with non-adaptive traits or mutations will be cleared away (sacrificed, you might say) for the good of the species. While many species regularly reproduce without sex, they do undergo sexual reproduction from time to time, taking advantage of genetic recombination. In experiments, sexless strains of the tiny nematode worm *Caenorhabditis elegans*, coevolving with their parasites, became extinct within twenty generations! Perhaps this is why sex is found in virtually all eukaryotes, from amoebas and redwoods to fish and fowl. Sex appears to be a necessary attribute for survival in a dynamic and unfriendly world.[17]

There are additional advantages to sex. One huge advantage is being diploid: having two sets of genes! In the diploid condition, you might be carrying a nonfunctional gene on chromosome number thirteen, but if your other chromosome number thirteen has a functional gene at the same position you're home free! Your good gene will be perfectly functional, "covering for" its nonfunctional partner. This has further advantages. Lots of individual genes differ only slightly from each other (called alleles). One allele may be advantageous in today's environment, but another allele may be more effective if the climate gets warmer. In effect, each of us diploid organisms carries *two genomes*—one from our male parent, the other from our female parent. Returning to the larger population, diploid species carry a large reservoir of diverse genes, all "held in waiting" for new environmental challenges. Sex, despite its costs, is ubiquitous among complex living things because it provides them a larger collection of genetic resources and keeps mixing them up. Though sex can be disastrous for individuals born with bad genes, it has proven essential for species-survival over evolutionary time.

COOPERATION AND BUILDING LARGER LIVING THINGS

Surely the invention of the eukaryotic cell, with all its internal components working together to carry on its life activities, was a major advance in the history of life. But that internal cooperation was followed in a few lineages by something even more extraordinary: the working together of many cells—cooperatively—to form larger multicellular entities. But how might unitary, independent cells, centered on their own reproduction, abandon those drives to form larger multicellular beings? Forming cooperative entities required that individual cells first develop mechanisms for *sticking together*. Next, they had to acquire *communication abilities* allowing them to interact with other cells and "do what they were supposed to do" on behalf of the larger organism. Such cooperating cells had to play by new and very restrictive rules.

Fundamentally, the basic purpose of all cells is to divide and make more cells. But that directive must be strongly modified if the cell is to be a functional member of a larger multicellular organism. In this case, the cell must take orders from its close neighbors. The tragedy of **cancer** begins when a few cells lose these constraints, proliferating at the expense of other tissues around them, and, finally, threatening the health of the entire individual. Cancer reminds us of how important the cooperation of cells is in maintaining the complex individual as a single well-functioning organism.

One of the simplest of animals, a sponge, has recently had its genome analyzed. Sponges are filter-feeders that lack symmetry. While very simple, their genome includes a basic genetic toolkit shared by all multicellular animals. Among these are genes responsible for some human cancers! These must be genes essential to forming the larger multicellular organism, instructing cells how to behave, and when to divide. Because they are multicellular, the lowly sponges require these genes just as we do. **Multicellularity** was one of the great advances in the elaboration of life on Earth. More remarkable, this advance in living complexity took place *independently* in animals, plants, and fungi! However, it did not come quickly. Larger multicellular organisms did not make their appearance in the fossil record until around 560 million years ago, a full four thousand million years after our solar system began!

We began this chapter focused on the most numerous and prolific of living things, the bacteria. From these simple cells, endosymbiotic unions allowed more voluminous cells to upgrade their energy processing. Utilizing the atmosphere's free oxygen, mitochondria gave the eukaryotic cell more energy more efficiently. Sequestered within a nucleus, the eukaryotic genome was protected from a metabolically turbulent cytoplasm. Meiosis and sex gave eukaryotic lineages the ability to reshuffle their genes to confront environmental change and to counter ever-diversifying pathogens and parasites. Together, these advances provided a platform from which eukaryotic cells would make new and innovative advances.

Chapter 3

WHAT DRIVES THE FORMATION OF NEW SPECIES?

In 1859, Darwin's treatise, *On the Origin of Species by Means of Natural Selection, or the Preservation of Favored Races in the Struggle for Life*, was published to great effect. This book presented a historical and naturalistic hypothesis for the perfection we see in nature all around us. Whether explaining colorful flowers attracting busy pollinators, or sleek antelopes evading hungry carnivores, Darwin's idea of natural selection over an immense period of time provided a reasonable alternative to specific deistic design.

In 1802, the Reverend William Paley had published *Natural Theology: Or Evidence of the Existence and Attributes of the Deity, Collected from the Appearance of Nature*. If we walk across a moorland and come upon a watch, Paley argued, we could easily see—from its complex inner workings—that this watch *had been designed* by the maker of the watch. And so, surveying a wide swath of natural history, Paley argued that the work of God's design was clearly evident all around us. Paley's book went through many editions; it was hugely popular and profoundly influential. Here was a richly documented treatise on what we now call **Intelligent Design!**

Nevertheless, and even if we were all to agree on an intelligent and prescient deity having created our world, an underlying question remains: how did the world around us actually come into being? Did God use his almighty powers to have the world's many wonders simply leap into existence from nothingness? Or did our God set the universe in dynamic motion (Big Bang!), decree the basic laws of nature, and then let the sweep of time fashion the world in which we live? Advocates of intelligent design have the same problem today; did new designs simply leap into existence out of nothingness, thanks to divine magic, or might they have come into being from simpler precursors through the workings of **Natural Law** over time? Here is where Darwin's book made its primary impact. By implying the operation of the same natural laws we see around us, Darwin's bold hypothesis opened the question of biological design to scientific inquiry.

Surely, if God used magic to fashion the things around us, we'd have little chance of understanding our world. We humans aren't much good at magic. But if much of the world has developed slowly over time by natural processes, then—like so many detectives—we can begin looking for the evidence of our origins. Darwin transformed the *design question* from one of Divine Mystery into one of pragmatic scientific inquiry.[1] Darwin coupled variation-within-species with selective survival and imagined these dynamics operating through deep time.[2] In fact, it was only with the realization that the Earth had a very long geological history that Darwin was able to postulate selection operating slowly over thousands of generations. His insight opened a new path for examining the historical development of the living world.

Unfortunately, there's a persisting problem with Darwin's book. *On the Origin of Species*, in fact, had very little to say about the origin of species. There are many pages about natural selection, variation in populations, breeding fancy pigeons, and lots of other stuff, but the differentiation of a population to become two new and distinct species gets only a few paragraphs. Even today, the creation of two species, where there was only one species before, is a controversial area in biological inquiry. We do know there's got to be lots of speciation on this planet. How else might we get to 380,000 species of beetles?

Let's confine our discussion to the question of how species actually become differentiated over short periods of time. If we look at the fossil record, there are places where that record is detailed enough to show how a single species has actually changed over many thousands of generations. The earlier representatives of this lineage may look quite different from those later formed and be given different species names. These are called chronospecies, but because the fossil record is highly fragmented, and there is always the possibility of a related species confounding things by wandering in from an outside territory, we won't trouble ourselves with speciation over evolutionary time. Instead, let's consider the factors that can cause species to split apart in contemporary landscapes, in what is called ecological time. But first, let's reconsider the species concept itself.

WHAT, EXACTLY, IS A SPECIES?

We have already touched on a species definition, but let's look at this concept more closely. If we go to Ernst Mayr's classic definition, we are told that species are "groups of actually or potentially interbreeding natural populations which are reproductively isolated from each other." What this clearly means is that species are on their own evolutionary track; they are not able to pick up genetic information from other closely related species. This concept demands that *interbreeding* and the exchange of genetic information occurs only within each species. From an abstract point of view this is fine: each species is an independent entity moving through evolutionary time.[3]

Unfortunately, Mayr's definition has some troublesome real-world problems. How do we know that the species we designate are not interchanging genes once in a while? We can't see or monitor "gene flow" in most wild populations. And the word *potentially* gives us botanists a big problem. Many plants that are able to interbreed when grown in a greenhouse do not appear to interbreed in the wild. Are these real species, or not? Botanists have tended to base their species concepts on what seems to be going on in the natural world. The word *potentially* was taken more seriously in zoology, where closely similar populations were often called subspecies—despite a lack of intermediates and evidence of gene exchange. Nowadays, with sophisticated DNA comparisons, ornithologists are finding that many "subspecies" are in fact genetically separate species. A recent study of two bird genera in Southeast Asia with ten accepted species and many subspecies has been transformed into sixty-one species! And it gets worse. The famous finches of the Galapagos islands—"Darwin's finches"—are beginning to look more like a dynamic hybrid mixture than an array of separate species with different lifestyles. Careful observation over many years, and through several El Niño cycles, points to a much more complex picture regarding Darwin's finches than was at first imagined.[4]

The situation in mammals is also troublesome. In 1982, Ian Tattersall revised the lemurs of Madagascar and recognized 36 species; by 2013, the listing of lemur species had escalated to 101 recognized species.[5] How much of this is based on new discoveries and better analyses, and how much of it is changing fashions in taxonomic practice? Clearly, the *species problem* is a continuing challenge.

A more fundamental concern is that we know very little regarding how genes move within and between populations of plants or animals living in the wild. Things are beginning to improve, thanks to new techniques of gene analysis. We had a pleasant surprise recently when Mary Ashley of the University of Illinois at Chicago and her students analyzed the paternity of acorns among bur oak (*Quercus macrocarpa*) trees here in Illinois. (Yes, DNA data can reveal acorn paternity!) Some of these bur oaks grow in stands isolated by open fields all around them—not an unusual situation in Illinois, where "oak groves" were often found surrounded by fire-swept prairies. With careful genetic analysis, each tree of an isolated oak stand was identified by its unique DNA profile, and it was revealed that a good percentage of the acorns produced in that stand had been "fathered" by pollen from distant trees outside the isolated stand![6] This was a surprise; most everyone had thought that virtually all the acorns within the stand would have been pollinated by trees within the same stand. In this case, wind pollination was much more effective over longer distances than we had thought. And, clearly, gene flow was more effective over longer distances as well.

In the case of animals, documenting gene exchange is also a problem. We are reminded of this when we find dead wildlife along our countryside highways, often during the breeding season when they wander more widely and get run over more often. Not just the animals, but the genes are doing a lot of "travelling" at this time. In addition, we do not know how effectively the northernmost populations are sharing genes with southern members of their same species. How then have we described more than 380,000 species of beetle species?

RECOGNIZING SPECIES

The classical approach to species recognition has been a pragmatic one: do they look different, do they behave differently, do they live in different habitats? Plant and animal species have been described on the basis of how they *differ from their closest relatives.* Even when they live together in the same habitat, can they be clearly separated by their morphological or behavioral features? If so, we call them different species. Thus, the "gaps" or differences between closely related species help us distinguish them.[7] Fundamentally, we assume that **species** look or behave differently because they are *not exchanging genetic information with each other.* Continuing gene exchange should cause a blurring of distinctions between different forms, and such varied forms cannot be considered different species. By using a combination of characters to separate two species, and finding that these differences are not bridged by intermediate individuals, we can conclude we are dealing with two different species.

Our own species is a fine example. Humans differ from each other within every population. In addition, there are continental patterns that are easy to see and supported by DNA surveys. While some regional groups of humans may be quite distinctive, we find intermediates with other groups at their borders, and all have the capacity to interbreed. Taxonomically, *Homo sapiens* is a single polymorphic species. (Though many anthropologists claim that "human races" are social constructs, the geography of DNA tells us otherwise.)

More significant with regard to the human species, there is no problem distinguishing any human with either of our nearest living relatives: the two species of chimpanzees. The variability of our own species is important; it reflects how human populations have adapted to local conditions over thousands of generations. Our varied ethnic differences, though minor, clearly reflect gene exchange over shorter and longer distances through thousands of years. However, there is no such thing as an animal intermediate between humans and chimps. There is a huge gap between both our physical and mental characteristics when we compare ourselves to chimps. Paleontologists estimate that this gap reflects a separation of more than five million years. Similar patterns can be found in many other animal and plant species. Each species may harbor considerable within-species variability, but with careful study we should be able to separate and distinguish them from their closest relatives.

When classifying plants and animals we are following in the footsteps of Carl Linnaeus, who initiated our system of **binomial nomenclature**. Simple, and used by scientists all around the world, the Latin binomial designates each and every species with two names. We call ourselves *Homo sapiens*. *Homo* is the name of the genus that includes living humans and a few extinct fossil representatives (such as *Homo neanderthalensis*). The *sapiens* part of our name is our specific epithet, our species name. Together, genus and species names form the binomial. The scientific name is italicized because it is in a foreign language: Latin. More important, species are the bottom rung of a hierarchy of Linnaean ranks. The genus is made up of species more closely related to each other than to species of other genera. Similarly, families are made up of genera more similar to each other than genera of other families. And so on up the line through orders, classes, and finally phyla (called divisions in plants). Though somewhat arbitrary, this hierarchy of ever larger groupings has proven immensely useful.

The binomial species name is, in fact, hugely informative. Having learned the name of the genus for the plant or animal you identify opens up an immense storehouse of knowledge. The genus will tell you the family, the order and the class to which your specimen belongs. Each of these ranks provides a broader array of features. In the case of *Homo* there is only one living species, but the genus name tells you that you are dealing with a primate, a mammal, and a vertebrate. All those categories carry information regarding the nature of the species. Our taxonomic system of ever larger ranks is a remarkably efficient way of organizing information about the world's living diversity. It is a logical system of **nested sets.** But let's get back to recognizing species.

Unfortunately, problems in recognition and separation are compounded when our so-called "species" seem to be exchanging genes. Eastern North American oaks (genus *Quercus*) are an annoying example. Finding a tree that looks like a **hybrid** between a bur oak (*Q. macrocarpa*) and a swamp white oak (*Q. bicolor*) isn't that uncommon. But their differences remain quite constant over much of eastern North America, and we continue to use these names as species. Are the hybrids contributing to gene flow? Perhaps, but hybrids may not be as successful in producing new offspring as are their parents. Because they are consistent over wide areas, we keep using our species concepts in oaks because they are useful in describing our native flora.

Another major problem in the species business, and one we'll gloss over, is that of **synonymy**—having more than one name for the same species. It stands to reason that taxonomist A in Austria and taxonomist B in Bolivia may discover the same species, whether in the jungle or in a museum collection. Each may realize that this species has not been described, and each may proceed to provide an appropriate scientific description and name. Publishing in different journals and at different times, there will now be two names for what may be the same species. But wait a minute! Careful comparisons of the type specimens (the specimens used to validate the names) may show minor differences. Are there two species here, or are these only slightly different variants of a single variable population? More collections and a study of their distribution should answer this question. But the problem remains: individual species may be burdened with many synonymous names, and this makes estimating actual species numbers a difficult task. Today, with the ascendancy of molecular biology, fewer taxonomists remain to get the names straightened out; as a result, many plants and animals have yet to be properly classified.

Coming back to Ernst Mayr's species definition, its importance is a philosophical one. The **Biological Species Concept** defines an independent lineage of plants or animals—independent because it cannot swap genes with other species. Such independence makes evolution a lot easier to explore. Best of all, the biological species concept works well for a high percentage of plants and animals. More significantly, Mayr's biological species concept informs our discussion of how new species form. By definition, the creation of new species—**speciation**—must be based on a cessation of gene flow between the new species and its parent species.

SPECIATION, PART 1: GEOGRAPHIC ISOLATION

The question of speciation—the formation of new species—revolves around the question of how gene exchange between populations can be terminated, and how that termination is rendered permanent. Larger scale geographical separation, surely, is the most obvious way to bring about genetic separation. If a rising mountain chain splits the range of a widespread species into two, and the separated populations are no longer able to interbreed, then slowly, over time, both will change. Each will pick up mutations not duplicated in the other, and each will adapt to the local circumstances within which they live. Both a "random walk" of new mutations and continuing adaptation to local conditions will cause the separated populations to go their separate ways. This process should, over time, result in genetic factors that prevent gene exchange, when and if the two separated populations have the opportunity to reunite later.

One can imagine the ancestors of horses and donkeys having become separated over many miles. Horses living in the steppes of Central Asia became expert at running over broad, relatively flat grasslands. Living in the rocky thorn bush of northeastern Africa and the Middle East, the ancestors of donkeys became less speedy but more sure-footed. Now, when mated, these two species form a very useful animal: the mule. Combining the greater size and strength of horses with the sure-footedness of donkeys, mules are very efficient beasts of burden. But there's a problem: they are sterile. The chromosomes of horses and donkeys do not line up properly in meiosis, and the sex cells of mules are not fully compatible. As a consequence, mules cannot reproduce! Adapting to differing environments, they became two species that are no longer capable of sharing genetic information—though they make a terrific hybrid. That's how geographic speciation is supposed to work.

Islands, surrounded by the sea, give us many instances of speciation by geographic isolation. Hawaii is the world's most isolated archipelago. Virtually all the native plants of Hawaii came from ancestors with minute wind-blown seeds, or they had fleshy fruit with seeds easily carried long distances by the birds that swallowed them. Today, over 95 percent of Hawaii's native plants are found nowhere else—that is, they are **endemic to Hawaii.** Likewise, the animals of distant islands had to have special attributes, allowing their ancestors to make the long journey over water. Arriving on an isolated island often presents new possibilities. A species of fruit fly (*Drosophila*) arrived in the Hawaiian Islands long ago; today these islands harbor almost a thousand species. The new colonists found unoccupied habitats free of their usual competitors, and diversified grandly. Lacking predation on an isolated island, some animals have become larger. Little turtles that rafted out to oceanic islands left their enemies behind, and they slowly became larger. The giant tortoises of the Galapagos are an example. But islands are often small, with not a lot to eat. Thousands of years ago, small elephants lived on islands in the Mediterranean, and in Indonesia as well. These small islands lacked larger predators and had a more limited food supply, resulting in selection for smaller, pony-sized elephants!

Another interesting pattern is the loss of flight among insects on small islands. Clearly, if you fly up off the ground and the wind blows you out to sea, you're not going to be contributing genes to future populations. On small windy islands, insects that cannot fly have a reproductive advantage, and they become the norm. Among the endemic species of carabid beetles in the Hawaiian Islands, 20 are fully winged but 184 have lost the ability to fly.

A fine example of "island isolation" has been documented among ferns growing on oceanic islands around South America. In this case the endemic island ferns—those found nowhere else—were shown to be related to *rare ferns* on the mainland. Apparently, it was easy to have common fern species arrive and become established on the island. But with continuing gene flow due to additional spores coming in from the mainland, these island plants "kept in touch" with their mainland conspecifics. With continuing genetic input, these island fern populations continued looking like their mainland counterparts. Since ferns can travel in the form of minute spores, they spread widely via the wind. New spores landing on the island will form new plants, and these will develop the sex cells that can interbreed with local ferns, maintaining genetic continuity with the mainland.

A different scenario played out when spores from ferns that were *rare* on the mainland arrived on the island. In this instance, it was less likely that the island fern would be joined by new immigrants from its mainland species. This new fern population was—in effect—more isolated on the island than populations of common species, and more likely to become distinctive. That's why the endemic island species proved to be related to rare mainland species, and not to the common ones.[8] Similarly, spiders of the Galapagos Islands are mostly the same species as those found in neighboring South America, while the spiders of the more distant Hawaiian islands include a great many endemic species.

Beginning in the 1970s, the idea of geographic isolation as a generator of new species became a topic of much discussion regarding the fauna of Amazonia. Jurgen Haffer, after studying bird distributions in the Amazon basin, came up with a fascinating scenario. During the colder parts of the ice ages, Haffer claimed, parts of the Amazon basin became drier grasslands, resulting in isolated patches of rain forest. Within these isolated patches, Haffer argued, new species of birds had recently developed, helping explain the present distribution of Amazonian forest birds.[9] Here, the island-like patches of rain forest had served as **refugia** in which the isolated forest birds could begin to form new species. This idea became very popular and engendered many concordant views. Trouble is, there was no real evidence that the Amazon basin had ever been dry enough to support extensive grasslands during recent glacial times. Pollen data did show that tree species from higher and cooler altitudes had shifted into the Amazonian lowlands during cool glacial periods. Indeed, the lowland rain forest had changed in species composition, but it had not been replaced by savannas to any extent.[10] Also, almost twenty cooler glacial cycles pulsing back and forth over the last two million years allowed precious little time for speciation-by-isolation. Though now discredited, Haffer's theory illustrates the popularity of seeing geographic isolation as a major generator of new species.

Clearly, geographic isolation is a factor in having populations splitting into new and separate species. But wait a minute; remember those 380,000 species of beetles? There simply aren't enough isolated mountains or islands to give us such extraordinary numbers. Also, hundreds of beetle species can live in the same small forest; some are even members of the same genus, living together as separate species. Consider human lice. There are two species; one lives in the hair on our head, and the other species lives in our crotch! That's not very far apart from a geographical point of view. How did this happen?

Worse yet, there is no guarantee that geographic isolation will result in genetic incompatibility. Sycamore trees (in the genus *Platanus*) from Eurasia and North America can produce perfectly fertile hybrids, despite having been separated geographically by the Atlantic Ocean over tens of millions of years. Hybridizing the eastern Mediterranean *Platanus orientalis* with the eastern North American *P. occidentalis* gives us vigorous hybrids used as street trees in many temperate zone cities. Not only are the parental species still inter-fertile, the hybrids are fertile as well, despite millions of years of separation. Such realities discredit any claim that long-term isolation will necessarily produce genetically incompatible new species. In a fine summary of speciation, Menno Schilthuizen concludes that "... scientific data are starting to support the notion that the impact of geographical isolation has been trivial rather than paramount."[11] Surely there must be ways, other than broad-scale geographic isolation, by which populations can form new species.

SPECIATION, PART 2: ECOLOGICAL DIVERGENCE

It was Alfred Russel Wallace, co-discoverer of the theory of natural selection, who first suggested that there might be a local pathway to species formation. Imagine a population living at the outer edge of the range in which most other members of its species are found. Imagine also that the *peripheral population* is in an environment somewhat different from the environment in which other members of its species are living. It stands to reason that this peripheral population will have to respond to environmental challenges different from those of its conspecifics. As such, the peripheral population will begin to adapt to its new habitat, slowly changing its genetic makeup. Wallace suggested that in such an instance, hybrid individuals—with genes from both the peripheral and the core areas—will be at a disadvantage! They will not be optimally adapted to the core area, nor will they be adapted to the peripheral area. In such circumstances, the hybrids should have decreased success, reducing gene flow between the peripheral population and the larger core species. "It is this inferiority of the hybrid offspring that is the essential point," wrote Wallace in 1883.[12] With one area of the species confronting one set of environmental challenges, while other conspecifics are subjected to a differing set of challenges, it makes sense that gene exchange between the two will be disadvantageous to both. Reduced "fitness" of the hybrids will, in effect, separate the peripheral population from its conspecifics, preventing gene flow between the two. Here is a way by which a species can *bud-off* small peripheral populations that, with luck, can become new and distinctive species in new and different environments. Maybe that's what happened on the human body, when one set of lice became adapted to a region of our body quite different from the region their fellow lice preferred.[13]

Early in my work with Costa Rica's flowering plants, I ran into a rather odd conundrum. The black pepper genus (*Piper*) had over three hundred species names ascribed to little Costa Rica, and hundreds of dried herbarium specimens needing to be identified. Most pipers are shrubs and subshrubs, easy to collect and well represented in museum collections. However, *experts* had created too many species names, and my job was to get the species numbers to conform with reality on the ground. Working with dried herbarium material easily separated distinctive species, but a crowd of specimens remained that were difficult to separate. The conundrum I encountered with *Piper* was this: though I had real problems trying to separate some of the dried specimens, in most any Costa Rican woodland even a child could easily distinguish the species. What was going on here? Over time, an answer presented itself: the most closely related—most similar—piper species were *not growing together!* The two "sister species" might grow on the very same mountainside, but *not* at the same altitude or in the same type of vegetation. Needless to say, these very similar species, flattened and dried, were the ones that gave me trouble at the museum. The altitudinal distinction in most of these close species pairs was easy to observe in the field, but much more difficult to evaluate in specimens. Later, working with many other Costa Rican species, I found this pattern repeated in other plant families. While not common, such close species pairs sent a clear message: they are the likely products of *recent ecological differentiation* on environmental gradients. In Costa Rica, species pairs separated on an altitudinal gradient were especially striking because they were all living in wet evergreen forests with similar rainfall. Yet one species would be found above 1,200 meters for example, and it's nearest relative at lower elevations. Most impressive of all, there were no intermediate (hybrid) specimens and no evidence for a gradual change up the mountainside (in what is called a cline). Clearly, it looked as if an ancestral species had given rise to a new species at the edge of its original habitat. This interpretation fits with the fact that these species appear to be each other's closest rela-

Recent DNA studies of species of cichlid fishes in Lake Victoria, East Africa, suggest that these hundreds of species arose from only a few introductions less than a million years ago. This should have alerted the scientific community to the possibility of speciation over very short distances. Further south along the Rift Valley, Lake Malawi has more than six hundred cichlid fish species, where a single introduction appears to have produced so many species.[16] Lake Tanganyika, located between Lakes Victoria and Malawi, has also had a radiation of cichlids. More important, the three different lakes possess species that have become specialized for similar niches! In other words, the separate lakes bear witness to independent and parallel adaptive speciation.[17] In each lake, cichlids gave rise to new species without the presence of significant geographic barriers.[18] Avoiding competition by swimming at different depths and pursuing different kinds of prey has created a richer fauna of ever more specialized fish.

The central point regarding ecological differentiation, as envisioned by Wallace, is that it provides an explanation for speciation over small distances: competition and natural selection. If gene flow from distant conspecifics *diminishes* local survivorship, then any mutation capable of reducing such gene flow will be selected for. In addition, if hybrids or intermediates suffer reduced fitness, they can act as a barrier to gene flow between the core population and its offshoot, just as Wallace suggested. The importance of these general concepts is that they function virtually anywhere on the planet, and continuously! Rising mountain chains, advancing glaciers, or diverging continents are unnecessary. All that is required is adaptive selection for a new and slightly different environment or new and different lifestyle. Continuing speciation, based on *environmentally driven divergent selection*, I believe, explains much of our world's extraordinary biodiversity.

The notion of ecological differentiation answers another question: Why should a genus have different species? Why not have gene exchange with all one's close relations? The answer would seem to be that in many cases it's better to be a specialist than a generalist. With differing species, a genus can have representatives of its lineage in a variety of different habitats or living different lifestyles. By becoming genetically isolated, these different species can adapt more effectively to their new habitats, free of the effects of nonadaptive genes from other close relatives. Such speciation scenarios may be more common in the tropics, where being an ecological specialist is often successful.

Further evidence for ecological speciation comes from the study of *Anolis* lizards in the Caribbean. Here, several lineages have produced similarly adapted species on different islands, and they've done this independently! Island isolation has given rise to parallel speciation within the same genus.[19] However, and with nature so versatile an enterprise, there are additional ways of forming new species.

SPECIATION, PART 3: OTHER WAYS OF MAKING NEW SPECIES

One of the simplest ways of creating a new species is through sudden chromosomal change. While rare among animals, this has been a significant process in the botanical world. Plants have messed around with chromosomes a lot, especially as regards the process called polyploidy. Recall that sex cells have one set of chromosomes; they are haploid. Most larger organisms are diploid, with each cell having two sets of chromosomes. Sexual union unites haploid gametes to form the diploid egg cell (zygote), which develops into an adult. But plants can do more. Once in a while they produce sex cells that do not divide properly and this, after fertilization, can create individual plants that have three sets of chromosomes (triploid), or four sets (tetraploid), or even more. **Polyploids** are individuals with more than the standard diploid number of chromosomes. Silly plants, they don't seem to realize something's wrong and often grow just as vigorously as normal members of their species. Better yet, here is a way of getting around "the mule problem."

Recall the mule: strong and sure-footed, but with chromosomes that just can't line up properly, rendering these animals sterile. What hybrid plants with differing chromosomes can do is to *double the entire set of their incompatible chromosomes!* Such a doubling may be a very rare event, but rare events in nature can have profound and long-lasting consequences. The individual hybrid offspring with doubled chromosomes experiences no problems during meiosis, because each chromosome now has an appropriate homologue to pair with. Suddenly, with doubled chromosomes, the once-sterile hybrid can produce functional pollen and egg cells. They can become, in effect, a new species. This is exactly what happened in the history of our most important cereal grain: wheat. Some many thousands of years ago, wheat's wild ancestor (einkorn) hybridized with a weedy relative. This may have happened often over time, but hybrid offspring could not persist. Enter the rare event: gametes that did not become reduced in chromosome numbers. When these got together in a new individual, we had a rare doubling of chromosomes. Bingo! Such was the origin of durum wheat, the stuff we use in making pasta; it is a tetraploid, with four sets of chromosomes. And as if all that weren't enough, durum wheat then hybridized with another relative, and BINGO again. This last episode gave us the bread wheats, with six sets of chromosomes (hexaploids). Certainly, ancient farmers tending their wheat fields had no understanding of what was really going on. But they did find and propagate both tetraploid and hexaploid wheat, foods that founded civilizations!

Though rare, chromosome doubling has been an important element in plant diversification. A more recent example involves American cord grass (*Spartina alterniflora*). Introduced into Britain in the late 1800s, this species hybridized with a native species there. The resulting hybrid was sterile and called *Spartina x townsendii*. Like its parents, the hybrid had sixty-two chromosomes, and the two pairs of thirty-one were incompatible. Nevertheless, these salt-marsh plants reproduced asexually, and the hybrids maintained themselves. Then, in 1892, a new and fertile cord grass was discovered; a hybrid plant had doubled its chromosomes to 124 and was now able to reproduce sexually. Called *Spartina anglica*, this new and invasive species has spread around the world! On a wider scale, recent DNA analysis suggests that chromosome doubling events played key roles in the early diversification of flowering plants.[20] Numbering over 280,000 living species, flowering plants bear witness to the efficacy of chromosomal change.

Animals, though doing little in the way of chromosome doubling, have many other ways of forming new species. Of these, new behaviors are among the most important. By developing a new and different song, or new and different plumage, populations of animals may be able to diverge and become reproductively isolated. Consider the peacock, with its grand show of fancy feathers. Darwin worried about peacocks; how might natural selection have resulted in so spectacular, but so burdensome, a display? His answer, and the one accepted ever since, was simple: choosy peahens! Apparently, the bigger the show, the more impressed are the ladies, and the more likely they are to mate. All this fuss and bother has an additional payoff: the females can judge the *quality* of their suitors. Darwin's **sexual selection** can explain fancy plumage, intricate mating rituals, and even provide new opportunities for speciation. Some of our eastern North American frog species, virtually identical in size, color, and shape, sing different songs at slightly different times. Their springtime songfest in swamps and wetlands is critical to bringing male and female frogs together for reproduction. Their different songs have made clear that these frogs are no longer exchanging genes, isolated by their springtime mating rituals. Perhaps they diverged after one population began singing a song just a little bit different from its neighbors. Along the same line, picture-winged fruit flies of the Hawaiian Islands are famous for having produced hundreds of endemic species; elaborate mating rituals seem to have been a key factor in their explosive diversification.[21]

WHY AREN'T THERE MORE SPECIES?

Here's a question that I find profoundly naive, a product of our modern, luxurious, famine-free lifestyles. In modern societies, we no longer lose a significant percentage of our children before they reach adulthood. We live in a world that no longer confronts the hazards most other species have to deal with. Before agriculture, we humans were constantly challenged by starvation, predators, parasites, and disease. The "real world" is different. Extinctions, local and global, are the rule, not the exception.

Imagining a world of dynamic equilibria where *bad things* happened on only rare occasions, many biologists of the 1970s dismissed the notion of "Nature red in tooth and claw!" But bad things happen often, and biological theorists have come back to their senses. Fierce and relentless competition in unpredictable environments not only created the incredibly diverse world we see around us, but continues to shape it. Species-loss and extinction events have occurred throughout the history of life; it is no small wonder that our planet supports as many species as it does.

Surely there must be many ways that population systems can become divided into genetically isolated species or bud off new sister-species. How else might the world have come to support so many species? And why have the great extinction events been followed by a resurgence of species numbers? Once an environment has been devastated, and many of its species sent into oblivion, the forces of renewal begin their work again. Empty environments invite new colonists, and once these are settled in, they seem impelled to further divide the landscape amongst themselves. Intense competition between members of the same species may be the underlying process. In this way, ecological differentiation provides a motive force for building ever greater species numbers. Forget about balanced ecological systems in some imagined equilibrium. Rather, speciation is dynamically driven by incessant competition, happenstance division, pest pressure, and new opportunities. Clearly, speciation has countered extinction effectively, making our world ever richer in species numbers.

Having a better notion of the ways in which new species can be generated, let's turn to the distribution of species numbers around the globe.

Chapter 4

THE GEOGRAPHY OF
SPECIES RICHNESS

Setting aside our focus on increasing complexity for now, we should remind ourselves of the splendid richness of life on land. In this chapter, we take a look at how terrestrial biodiversity is distributed around the planet, and in the next chapter we explore significant patterns in those distributions. Certainly biodiversity is not evenly distributed around the globe.[1] Some regions are graced with a multitude of species, while other areas have few visible residents. Terrestrial species numbers are determined by two primary factors: temperature and moisture.

Where temperatures are warm and rainfall is plentiful, biological diversity reaches its full expression. Where temperatures are frigid or moisture is scarce we find wind-swept ice fields and barren desert. Extreme conditions constrain the exuberance of life and have a global signature. Our planet is marked by several distinctive climatic zones.

At the poles, where sunlight is less direct, freezing temperatures predominate. The South Pole is in the center of a single frigid continent—Antarctica—surrounded by the turbulent Southern Ocean. In contrast, the North Pole is at the center of an ocean, much of it frozen. Neither pole provides a friendly environment for complex terrestrial life, though the cold waters do support a wealth of marine creatures. Our planet's polar biotas are antipodal, widely separated by temperate and tropical zones. In the natural world, polar bears, emblematic of the Arctic, have never encountered wild penguins, confined to southern seas.

The world's most severe deserts also have a global pattern; they are clustered around the Tropic of Cancer (at 23.5° north latitude) and the Tropic of Capricorn (at 23.5° south latitude). Here, longer days are coupled with intense sunshine during the summer months. This is where Earth's highest temperatures are often recorded. Tropical rains do not reach these latitudes, and zones of high atmospheric pressure, hovering around 30° north and 30° south latitudes, result in broad bands of limited rainfall. The world's largest desert region ranges across the entire northern tier of Africa (the Sahara), continuing eastward across Arabia into western India. Similarly, severe dry areas are found in the subtropics of the Southern Hemisphere, especially in southern Africa and across central Australia.

In addition to polar frigidity and subtropical belts of aridity, our planet features broad zones of tropical and "temperate" vegetation. All told, our planet supports many different biomes, all contributing to the variety of species richness around the globe.[2] Unfortunately, defining and delimiting different biomes or vegetation types is often a matter of convenience: more art than science. As rainfall diminishes or temperatures change across hundreds of miles, local floras change gradually in stature and species composition. Distinguishing North America's tall-grass prairie from the short-grass prairie or semi-desert steppe can be quite arbitrary. These biomes all intergrade, and their varied habitats share many species of plants and animals as well. It is only over longer distances or over sharp changes in elevation that one can clearly see that the vegetation has changed. However arbitrary, dividing the natural world into biogeographic realms, biotic provinces, ecoregions, biomes, or plant communities helps us understand the natural world. Our land surfaces have been divided into as many as 867 ecoregion units.[3] Similarly, 426 freshwater ecoregions have been delineated.[4] Though useful on a local scale, such smaller units are too detailed an inventory for our brief survey. On the other hand, zoological geographers have recognized as few as six major faunal realms around the world: too broad a brush for our discussion. Instead, let's survey the larger geographical regions by their major vegetation zones. Since green plants capture the energy of sunlight and provide much of the community's physical structure, vegetation is an appropriate focus for our overview.

More significantly, vegetation is usually the primary factor in determining overall regional biodiversity, both in terms of numbers of species and variety of lineages.[5] For example, a recent study found that the high numbers of herbivorous insects in tropical rain forests is due to the high number of different plant species, not to more restricted feeding patterns on the part of the insects.[6] Across Australia, bird diversity is closely correlated with vegetation and evapotranspiration rates (a measure of water loss for plants).[7] Examining vegetation around the world, let's begin at the "top" of the world, where the numbers are low.

THE FRIGID NORTH

Modern cartographers place the North Pole at the top of their maps, and from this perspective we see that a large majority of our planet's land surfaces are located in the Northern Hemisphere. Both the Eastern and Western hemispheres have huge land areas bordering the Arctic Ocean. In dramatic contrast, the Southern Hemisphere has wide oceans covering much of its breadth, and harbors many isolated islands as well. Thus, with the great majority of land surfaces at the "top" of our planet, we begin our journey in the north.

Bordering the frigid Arctic Ocean, we find a circumpolar belt of vegetation called the **tundra**. Ranging from a few centimeters to less than a meter (3.3 ft.) in height, this vegetation is composed of hardy and wide-ranging plant species, but there are no trees. Northernmost Russia differs little in its species listings from northernmost Canada and Alaska. Here, many species of mosses, liverworts, and lichens contribute to the ground cover. They survive the harshest conditions, on soils permanently frozen just a few inches below the surface. Members of the blueberry family, sedges, fescue grass, and a number of colorful flowers are significant elements of the tundra vegetation. In more protected sites, stunted little shrublets of willow, larch, spruce, and alder can be found. Because of the climate's severity, tundra plants are perennials, hanging on to the landscape by the growth of previous years. With a growing season of less than fifty days, flowers often bloom in one year and form their seeds the following summer. With the unfortunate exception of blood-sucking mosquitoes, insects are few. Larger mammals of the tundra are the caribou (called reindeer in Eurasia), hares, a few rodents, and arctic foxes. Small passerine birds are few, with larger, warmly insulated geese, ptarmigans, and owls active during the summer. Larger land animals survive the long winter by hibernating or migrating southward into the boreal forest. In contrast, many larger birds, seals, walrus, and polar bears live along the edge of the cold northern sea, feeding on marine life. They border but are not part of the tundra ecosystem.

Surely, one the most *fortunate* qualities of our planet is that it spins at a slight angle in respect to the plane of our orbit around the Sun. Consequently, summery days in the Arctic north may be few in number but long in hours. Without this extended summertime sunshine, land bordering the Arctic Ocean could not support a tundra vegetation. But the winters are long and cold, with soils thawing only a few inches above the permafrost in summertime. And because roots are unable to absorb moisture from a frozen substrate, trees cannot sustain evaporation from their leaves in this environment.

Overall, the northern tundra ecosystem supports fewer than one thousand species of higher plants. That's not a lot of biodiversity over almost one-fifth of the world's land surface. We'll be using higher plants or flowering plant numbers in many of our comparisons. (Higher plants—vascular plants—have an internal plumbing system. They include flowering plants, conifers, and other seed plants, as well as ferns and fern allies. The "lower," nonvascular plants include mosses and liverworts, while lichens are part of the fungal kingdom.) Importantly, tundra-like vegetation is not limited to polar areas. Even in the tropics, high wind-swept mountain tops support tundra-like meadows, alpine grasslands, and rocky surfaces decorated only with lichens and mosses.

NORTHERN (BOREAL) FORESTS

Leaving the tundra and moving south, we enter the northern **boreal** forest. This biome is also called the northern conifer forest, evergreen needle-leaf forest, or by the Russian designation *taiga*. One of the largest ecosystems on our planet, boreal forest covers about 29 percent of the world's forested land.[8] Toward the north, scattered little trees intergrade with the tundra, becoming more numerous and taller toward the south. With soils often frozen only a few feet below the surface, drainage is limited and lakes and bogs are common. Cone-bearing trees with needle-like leaves dominate the boreal forest; these include the spruces (*Picea*), firs (*Abies*), pines (*Pinus*), and larches (*Larix*). Slender conical shapes help these trees survive long fierce winters, as wide lower branches brace the short upper branches when burdened with ice and snow. Flowering plants are represented among the trees by willows (*Salix*), birches (*Betula*), aspen (*Populus*), and alders (*Alnus*). Members of the rose, blueberry, composite, and other families contribute to the shrubby layer. Grasses and sedges, as well as mosses, liverworts, and lichens, are significant in the ground cover. Lichens, in fact, are a primary food for the caribou here, as they are in the tundra. (We'll discuss lichens in chapter 6). Conifer seeds nourish voles, lemmings, and a variety of birds. It has been estimated that these extensive forests of the north include a third of all the world's trees. But that's a tree count, not a species count. These cooler biomes of the north are not rich in species. Alaska and the adjacent Yukon are home to about 1,560 vascular plants, while Norway numbers 1,715. (These numbers include species of both the tundra and the boreal forest.)

Though similar in overall form and stature, the boreal forests of North America and northern Eurasia differ slightly in species composition. Among larger mammals, the populations may differ only enough to be called subspecies. Both regions are home to wolves, foxes, beavers, elk, lynx, badgers, bears, and the world's largest deer: the moose. Because of this, the northern boreal forest zone is usually divided into two: the American, or Nearctic, and the Eurasian, or Palearctic realms. Biogeographic terms like Nearctic and Palearctic reflect the impact of Columbus's discovery; for Europeans the Americas were new and unexpected! Traveling further south, we encounter more temperate climates and a vegetation much richer in species.

THE NORTHERN TEMPERATE ZONE

Below about 50° north latitude, we find "temperate" regions encompassing a great variety of biomes over a large portion of the world's land surfaces. In North America, and beginning in the west, we find dense evergreen conifer forests bordering the Pacific coast, ranging from Alaska to northern California, where we find the world's tallest trees. In fact redwood forests may comprise as much as 3,500 tons of biomass per hectare (2.47 acres), fully four times as weighty as a tropical rain forest. And though these coastal conifer forests are resplendent in lofty tree trunks, they lack the biodiversity found in many other temperate forests.

Moving eastward from the coast, we encounter the results of massive tectonic convulsion: the sharply sculpted sierras, high intermountain basins, and older Rocky Mountains. Beginning in southern Alaska and northwestern Canada, this rugged topography extends southward into western Mexico. Conifer forests and grasslands dominate these rugged landscapes, thanks to limited rainfall, higher elevations, and frequent fires. Moving farther eastward, the sharply rising Rocky Mountains mark the western edge of North America's broad interior plains. Ranging from Alberta in the north to Texas in the south, the Great Plains experience restricted rainfall, frigid winters, and hot summers. Before their slaughter in the late nineteenth century, sixty million bison roamed North America's broad interior plains, in addition to mule deer, elk, and the fleet pronghorn antelope. Drier in the west and moister toward the east, the Great Plains support short-grass **steppe** in the west and tall-grass **prairies** in the east. Partly as a result of fire, trees are often restricted to river valleys and moist protected slopes. By setting fires regularly, Native Americans expanded tall-grass prairies eastward into more densely forested Illinois and Indiana.

With cold winds from the north and warm moist air from the Gulf of Mexico, North America's central plains are subjected to a rare weather phenomenon: tornadoes. When warm moist air from the Gulf of Mexico clashes with cold dry air coming down from Canada, we have the possibility of storms with limited area but horrendous power. America's central plains have something no other place on Earth can match— "Tornado Alley"!

Eastern North America, with more reliable rainfall, milder temperatures, and the Appalachian Mountains, supports evergreen coniferous forest, broadleaf deciduous forest, and mixtures of the two. Such woodlands once carpeted the region from the western Great Lakes southward to the Gulf of Mexico, and from eastern Canada to Florida. Ecologists recognize about ten different forest assemblages in this region.[9] The flora is especially rich in the southern Appalachians, home to the world's highest concentration of salamander species! Again, though we call them "temperate," the eastern United States are subject to long and frigid winters. Only southern Florida escapes freezing temperatures, thanks to the Atlantic's nearby Gulf Stream.

For an overview of species richness, the multi-volume *Flora of North America* covers all the seed plant species of North America, north of Mexico. All biomes are included, from Arctic tundra, evergreen conifer forests, and deciduous forests, to grasslands and deserts. Overall, this flora numbers about 19,500 species of native vascular plants. As one would expect over so vast an area, species richness varies greatly. The richest single biome of this large region is the broadleaf deciduous forest of the eastern Appalachians. The northeastern United States, from east of the Great Plains and north of Virginia to southern Canada, numbers 4,500 species of higher plants. North and South Carolina have around 3,500 native species (together with about 1,000 introductions). In contrast, and with a fifth of the area of all lower forty-eight states, our drier interior Great Plains number only around 3,100 species of native plants. California, with a great variety of biomes over a large and mountainous topography, numbers about 5,700 species of higher plants. Differing patterns of rainfall and altitude allow California to support vegetation ranging from seasonally parched chaparral in the south to tall evergreen conifer forests along the north coast, and a variety of montane and sub-desert vegetation as well. Here live the world's largest living things—Giant Sequoias—and the oldest—Bristlecone Pines, some more than four thousand years in age.

Despite a fine diversity of habitats and large areas, North America's biodiversity does not compare well with species numbers in the tropics. Returning to the Midwest again, but shifting to butterflies, we find that the state of Michigan supports only 134 species of butterflies over its 150,780 km^2. Compare that to tropical Panama, which numbers 1,550 species of butterflies over its 78,200 km^2 area. Clearly the temperate zone is a not an easy place to make a living. Remaining at similar latitudes, we'll move to the world's largest temperate region: Eurasia.

NORTHERN EURASIAN BIOMES

The world's largest single land area covers much of the Eastern hemisphere's northern half; arbitrarily separated into Europe and Asia. Northern Europe's biotas are dramatically separated from their southern neighbors by the Pyrenees, Alps, and Balkans. These mountain chains—running west to east—inhibit interchange between Europe's northern flora and the drier and warmer Mediterranean vegetation to the south. France's flora of higher plants numbers around 4,600 species, while more northerly Germany has only 2,700 species. Though poorer in species than its eastern American equivalent, the flora of northern Europe is dominated by many of the same genera, such as oaks, maples, birches, and a variety of conifers. Europe's deciduous forests intergrade with boreal forest, both to the north and eastward across Russia.[10]

Western and central Asia's huge interior is also bounded along its southern edge by mountains, from the Caucasus to the Himalayas. Situated between the Black and Caspian Seas, the Caucasus region harbors a robust 6,300 species of vascular plants, with 1,600 of them found nowhere else. These high numbers have led to this region being designated a "biodiversity hotspot."[11] (We will discuss "hotspots" in the next chapter). Western Asia's broad interior supports montane conifer and deciduous forests in moister areas, and grasslands intergrading with semi-desert steppes in the high plateau north of the Himalayas. Unlike North America, western Asia's interior is drier to the south and moister to the north.

Resembling eastern North America, eastern Asia has more plentiful rainfall and was once covered with broad-leaved deciduous and evergreen forests. Higher elevations, from the Himalayas into China, are dominated by conifer forests. Eastern Asia's broad-leaved forests once ranged, uninterrupted, from the boreal north to the tropical forests of Southeast Asia. Despite dense human populations, China supports an estimated flora of 32,870 species.[12] This number far exceeds the 19,500 species in all of North America north of Mexico! Varied and complex mountains in southern China and the eastern Himalayas are part of the reason, but history has also played a role.

Why might China be so much richer in species and genera of vascular plants than either eastern North America or Europe? The answer seems to involve recurrent glacial advances over the last two million years (the **ice ages** or Pleistocene epoch), coupled with each region's specific topography. While the Himalayas are oriented east-west, a majority of mountains in southwestern China are oriented north-south. Here, plants and animals could migrate in response to warming or cooling climates, north or south. Simply stated, eastern Asia suffered less extinction than occurred in the floras of northern Europe and northeastern America. In these two regions, plant migration was restricted by the mountains of southern Europe, and the Gulf of Mexico; today's species numbers reflect that history.

THE WORLD'S "MEDITERRANEAN" FLORAS

In contrast to its northern neighbors, the floras of southern Europe, the Mediterranean area, and southwestern Asia are unusually rich in species. This **Mediterranean Flora** has hot and dry summers, followed by a dry fall. Rains begin during the cold of winter and extend into spring. As temperatures rise in springtime, a majority of plants bloom together in one grand spectacle. Half the plant species are annuals, while shrubs and small trees bearing small stiff evergreen leaves dominate the landscape. In addition, highly dissected mountains provide a wide range of habitats around the Mediterranean Sea and southwestern Asia. In spite of limited rainfall, the Mediterranean flora is estimated to harbor 25,000 vascular plant species, a number that easily exceeds the flora of all of North America!

Greece, though relatively small in area, is home to about 5,000 species of higher plants, while larger Turkey numbers around 8,600 species. Complex mountain topography and isolated islands are major factors in sustaining biological diversity in this part of the world. However, the Mediterranean and Middle East are not the only parts of the world to support this special biome.

Our planet is graced with four additional "Mediterranean floras." All are subtropical, located between twenty and thirty degrees from the equator, and mostly found near the western flanks of large continental land masses. In addition to the Mediterranean, they occur in southern California, central Chile, southwestern Australia, and at the southern tip of Africa. All are open woodlands or shrub lands with drought-tolerant plant communities. Summer and fall are hot and dry, while a colder winter and early spring bring the rainy period. And, with a long dry season, all are fire prone. Despite being widely distant, these biomes have a number of similarities. With limited rain fall, animal life is not especially rich. However, there are unusual similarities in the faunas. Southern California's chaparral supports 235 species of non-marine birds, while the *matorral* of central Chile supports 230 non-marine bird species; a striking similarity.

Each of the Mediterranean floras is famous for the spectacle of their springtime flowering. In late September, five-day tours allow tourists to survey wildflowers near Perth, Western Australia. Current estimates claim that this Mediterranean-type flora numbers 7,380 native species of vascular plants. That's a terrific number for so dry a corner of the world. South Africa's Mediterranean-type flora is home to an incredible 9,000 native vascular plant species (something we'll come back to in the next chapter). More generally, the world's Mediterranean floras are home to about 48,000 species of higher plants. That's close to 20 percent of the world's higher plant species on only 5 percent of its land surface and explains why these floras have given us so many useful food sources and ornamental plants![13]

THE AMERICAN TROPICS

Excepting high mountain-tops, life in the tropics does not experience the challenge of a killing frost. Here, with sufficient moisture, species reach their greatest numbers. Biogeographers refer to the American tropics as the Neotropics (again, referring back to a world that was new for Europeans). Moving from north to south, the Neotropics begin in Mexico. Rich in floristic diversity and with complex topography, Mexico has thirty-two distinctive vegetation types, supporting an estimated 26,000 species of higher plants.[14]

Central America, embroidered with complex mountains and many volcanoes, is flanked by a moist Caribbean slope on the east, and a seasonally dry Pacific coast. We botanists define the Mesoamerican floristic region as ranging from the narrowed Isthmus of Tehuantepec in Mexico, south to the Panama-Colombia border. With about 17,000 species of flowering plants, the Mesoamerican flora approximates the species numbers for all of North America north of Mexico.

In designating Mesoamerica a "'biodiversity hotspot" Russel Mittermeier and his associates estimated a fauna of 2,859 non-fish vertebrate species in this region, with 1,159 (40.5 percent) found nowhere else.[15] Reptiles contribute 685 species to these totals. These many species live in a variety of Mesoamerican biomes, ranging from seasonally dry deciduous woodlands and highland conifer forests to true lowland rainforests, montane evergreen forests, and high montane alpine formations.

SOUTH AMERICA

South America supports the world's most biodiverse region. Running along this continent's entire western edge are the Andes Mountains. Not only the largest mountain range to be found anywhere in the world, this system runs from north to south and across the equator. Thanks to the north-south orientation, cold-requiring plants and animals were able to migrate along the Andes' highest elevations and survive through glacial cycles. Even more important, the Andes' grand variety of elevations and complex topography affords many varying habitats; allowing Colombia, Ecuador, and Peru to support floras of more than 16,000 species of vascular plants each. The number of recorded bird species in Colombia and in Peru is around 1,800, more than twice the 770 recorded for all of the United States and Canada.

After the Andes, the Amazon basin is the second most obvious physiographic feature of South America. Covering 7,050,000 km² (2,722,000 sq. mi.), the Amazon drains about one-fourth of the continent. With over a thousand tributaries, the world's largest river system contains about one-fifth of the world's flowing fresh water. The overall number of freshwater fish species for the Amazon basin is estimated to be 3,000, more species than live in the Atlantic ocean! Despite its vastness, the basin is relatively flat. Combining low relief with strongly seasonal rainfall, some places in the basin have river levels rising and falling as much as ten meters (thirty-three feet) each year! These annual cycles result in large areas of seasonally flooded forest (*varzea*) and continuously flooded swamp forest (*igapo*). The highest species numbers, however, are found in the non-flooded forests on higher ground (*terra firme*)—especially along the eastern slopes of the Andes Mountains.

Other areas of South America support distinctive floras as well. The northeastern Guiana Shield is marked by isolated table mountains harboring many unique species. In Brazil, dry scrubland in the northeast, remnant evergreen forests along the southern Atlantic coast, and a rich deciduous vegetation in the south support many endemic species. These floras, together with the Amazonian basin, give Brazil an estimated 56,000 higher plant species! And that number keeps growing; 2,875 new flowering plant species were described from Brazil between 1990 and 2006![16] The *cerrado*, a seasonally deciduous woodland, is especially rich in both species numbers and endemics. Around 6,400 higher plant species grow in this biome, together with 837 species of birds and 1,000 species of butterflies. Endangered by development, Brazil's *cerrado* has been designated another biodiversity hotspot.[17] Grasslands are also important to South America's species richness, especially in the Orinoco basin, in the Andean highlands, and over much of Argentina. Cooler temperate forests at the southern tip of the continent and coastal Pacific deserts add other distinctive biomes.

Regional ecological provinces across South America can be divided in a variety of ways, with the Andes and lowland Amazonia the largest. However, the Andes can be subdivided along the north-south axis, near the Ecuador/Peruvian border, where there is a depression in the mountains. A more important division is elevation itself: above about one thousand meters elevation, moist eastern slopes support montane cloud forests. On the far drier Pacific slope, subdesert, grassland, and scrub abound. The highest elevations, between three thousand and five thousand meters, have their own specific biomes; here is where the potato was first cultivated. All together, these many biomes over huge areas give South America its high species numbers.[18]

TROPICAL SOUTHEAST ASIA AND NEARBY ISLANDS

As in the Neotropical world, the Paleotropics contribute greatly to the world's biodiversity. Biomes in India range from parched desert to evergreen rain forests, with the subcontinent numbering about 25,000 higher plant species. Southeast Asia tends to have greater rainfall, and this region includes many island archipelagos. Malaysia, a long narrow peninsula, is home to about 15,500 species of higher plants. The island of Borneo has about 13,000 flowering plant species, with Mount Kinabalu (4,101 m; 13,455 ft.) supporting many endemic species.

Not only are the rain forests of Southeast Asia rich in species, some are unusual in physical structure and composition. Many of these forests are dominated by a single family of flowering trees, the Dipterocarpaceae. These trees have two unusual features. The first is well-known: unpredictable flowering after several years of not producing any flowers or fruits. Not only do a variety of species and genera join together in these unpredictable flowering extravaganzas, *they fruit synchronously*! (El Niño weather oscillations may be the cause of these unpredictable cycles). The resulting pattern of huge fruit production, followed by several years with little to eat, is a challenging environment. Consequently, the forests of this part of the world do not have quite as many animals living on the forest floor as do Neotropical rain forests.

A second feature of these Dipterocarp forests is the height of their trees. American rain forests may have emergents reaching 50 meters (165 feet) in height, but rarely do they reach higher. In contrast, trees reaching 70 m (230 feet) are not uncommon in the forests of Southeast Asia. The dipterocarps, with their slender trunks and high crowns, give the forest canopy a tall and more open, spindly, appearance. Local animals have responded to this unique canopy architecture. These are the only forests where you can find flying frogs, flying snakes, and a flying lizard. Of course, they're not really flying; these creatures *glide from tree-to-tree* with flaps of skin along bodies and tails and webbed fingers. No other forests in the world support such a variety of gliding animals. What puzzled me was that these tall slender trunks carried a crown of relatively few leaves. A recent study explains this phenomenon: Dipterocarps appear to photosynthesize more efficiently than trees in other families; they simply don't need as many leaves!

Also, there is history! Two remarkably different faunas abut here, along what is called "Wallace's Line." Running roughly from the Indonesian island of Lombok northward between Borneo and the Celebes, this imaginary line has animals such as elephants, deer, leopards, and monkeys on its western side, and marsupials, a greater number of lizards, and birds of paradise to the east. Zoogeographers use Wallace's Line to separate the **Oriental Realm** from the **Australian Realm**. (We'll analyze "Wallace's Line" in the next chapter.)

AUSTRALIA: A WORLD APART

Australia is an odd continent with a strange biota. Just about flat, only 5 percent of its surface rises more than 100 m (328 ft.) above sea level. The lack of recent geological activity—there are no volcanoes—has resulted in weathered nutrient-poor soils. In addition, recent tectonic displacement has shifted Australia from a moister, more southerly location northward into hotter and drier subtropical latitudes. Fossils indicate that Australia has lost much of its earlier flora, with severe drying over the last twenty-five million years. Today, two-thirds of Australia has an annual rainfall of less than 500 mm (19.5 in.). Poor soils, together with unpredictable rains and frequent fires, make this the world's least productive continent.[19]

Isolated in the Southern Ocean over many millions of years, island Australia's fauna lacks the more advanced placental mammals, such as monkeys (primates), deer (ungulates), and cats (carnivores). (Dog-like dingoes arrived more recently with people.) Not only does Australia sustain marsupials, such as kangaroos, wallabies, wombats, and koalas, it harbors an even more ancient mammalian lineage: the monotremes. With skeletal features resembling reptiles, these are the only living mammals giving birth by laying leathery eggs! The semi-aquatic duck-billed platypus (eastern Australia) and several species of spiny echidnas (Australia and New Guinea) are the only survivors of this ancient group. In addition, the bird and reptile faunas are rich in species. All told, Australia has the most unusual land vertebrate fauna in all the world.

Despite the dry conditions and poor soils, Australia's higher plants number around 15,000 species over an area of about 7,899,850 km². Many landscapes are dominated by trees in the genera *Eucalyptus*, *Acacia*, and *Casuarina*—trees and shrubs that tend to remain green even through severe drought. Intense fires have resulted in many woody plants having seeds protected by thick woody structures. *Hakea*, a genus of woody shrubs, has solid woody fruits about the size of a walnut that split down the middle to release a single winged seed. *Banksia* inflorescences are thick colorful, upright spikes with many flowers densely arrayed around a columnar center. That central axis becomes a thick woody cone, with seeds developing inside; odd, lip-like openings on the surface provide exit for the deeply hidden seeds. *Eucalyptus* trees protect their seeds within a little urn-like woody fruit, topped with a circular lid. These woody fruits all open after a hot burn to release the protected seeds. Concordant with both fire and drought, Australia has few trees with fleshy fruit and, consequently, very few frugivorous birds or mammals. Australia, however, is not the only continent that has been sharply affected by drought.

AFRICA SOUTH OF THE SAHARA

Bordered along its north by both the Sahara and Arabian deserts, biogeographers delineate the African realm as **Africa South of the Sahara**. Celebrated for a rich and spectacular mammalian fauna, Africa is a continent well-endowed in biodiversity. Here are the world's only grass savannas where large grazing animals still roam freely by the thousands. The Serengeti plains have been estimated to support four million zebras, wildebeests, and gazelles, with about three thousand lions culling them. Rich rain forests persist along coastal West Africa and over the Congo River's wide basin. Our closest relatives—chimps, bonobos, and gorillas—still live in some of these moist evergreen forests.

Eastern Africa differs greatly from the western half; it is marked by the Great Rift Valley system, one of our planet's most distinctive geological features. Beginning in Mozambique, this system of elevated highlands and north-south oriented valleys extends northward through Tanzania, Kenya, Uganda, and Ethiopia into the Red Sea. Geological uplift and volcanism along the Rift have given eastern Africa diverse topography, rich soils, and a grand variety of habitats, ranging from lowland semi-deserts and grasslands to acacia woodlands and higher mountains bedecked with evergreen forests. It was in this rich and diverse landscape where a two-legged primate became human.

Africa's fauna includes many striking endemic mammals, such as hippopotamuses, giraffes, elephant shrews, naked mole rats, and bush babies (small nocturnal primates). The pig-sized and termite-eating aardvark once ranged over all sub-Saharan Africa and resembles nothing else. Colorful and often large, the birds of Africa are another glorious aspect of the fauna. However, and in strong contrast to the fauna, Africa's tropical flora is relatively poor in species numbers.

When counting plant species, tropical Africa doesn't measure up to either the American tropics or tropical Asia. For example, little Singapore island, at the end of the Malay Peninsula, is home to eighteen genera and fifty-one species of native palms. That's close to the fifteen genera and seventy-two species of palms native to all of Africa! The native higher plants of Ethiopia probably number around 7,000 species, despite Ethiopia's ample area (at 1,127,130 km^2 it's bigger than Texas). Though twenty times the area, Ethiopia's species numbers are far fewer than those of little Costa Rica, with around 10,600 species on about 51,100 km^2. Both countries are tropical, sit about ten degrees north of the equator, get their primary rains between April and October, and both have elevated montane habitats. Yet Ethiopia doesn't come close to Costa Rica's floristic richness. An astounding 1,300 species of orchids have been found in little Costa Rica. In contrast, Ethiopia's native flora contains fewer than 200 orchid species!

On the western side of Africa, the flora of Nigeria, with an area of about 923,770 km^2, is home to only 4,715 higher plant species—again, much less than little Costa Rica. The Democratic Republic of Congo, with an area of 2,345,410 km^2 and vast areas of rain forest, sustains a flora estimated at around 11,000 species. A few animal lineages are also poorer in Africa than in tropical Asia and the Americas. Comparing the butterflies of three tropical countries with similar areas, we find that Liberia in western Africa has 720 species, the Malay Peninsula has 1,031, while Panama has 1,550 species. Liberia with 111,370 km^2 has fewer than half the butterfly species of Panama with about 78,200 km^2.[20] These are significant differences!

However, species numbers are very different at the southern end of Africa, where a more temperate flora numbers an astonishing 23,400 higher plant species.[21] What can explain why the huge central core of Africa—tropical Africa—numbers only around 27,000 species?[22] This is less than the flora of China and similar to individual countries in the northern Andes, all with smaller areas. Why is Africa's tropical region so poor in plant species?

Several factors may have played a role in reduced floristic species richness in Africa's tropical midsection. First of all, there are no large mountain chains in tropical Africa, although areas along the eastern rift are elevated. Also, central Africa is bordered with many harsh deserts. The Sahara sweeps across the entire northern fifth of the continent. Sudan, with 2,505,800 km^2, has a flora of only around 3,200 native plant species. The Namib and Kalahari are large deserts covering much of southwestern Africa. In addition, even moister areas of Africa are often seasonally stressed for rainfall. Nowhere in Ethiopia does the rain fall as reliably and as generously as it does on the Caribbean slopes of little Costa Rica. However, the primary reason for tropical Africa's paucity of plant species is most likely historical. Africa probably suffered one or several crises of drought in the not-too-distant past.[23] Also, it is still dry. Annual leaf-area indices—as measured from satellites—are much lower for Africa than they are for South America. Consider those African plant genera that are rich in species numbers; they are grasses, acacias, cactuslike euphorbias, and members of the hardy thorn-bush flora. An unusual aspect of Africa's flora is that it is not particularly poor in families or genera of flowering plants, compared to the American and Asian tropics, suggesting that widespread drought decimated the vegetation, leaving fewer species but having a lesser effect on higher ranks of plants.[24]

Biogeographers once placed Africa-south-of-the-Sahara together with Madagascar into a single "Ethiopian realm." However, the biota of the island of Madagascar is so unusual that it is now considered its own zoogeographic realm.

MADAGASCAR AND OTHER ISLAND BIOTAS

Moored just east of southern Africa, in the Indian Ocean, we find one the world's most distinctive biotas. Island Madagascar is the home of an endemic and ancient primate lineage: the lemurs. With pointed dog-like faces, these monkey relatives resemble fossil primates from earlier times. Despite being close to Africa, this island has not a single ungulate species (pigs, antelopes, and their kin). Likewise, there are no true cats, no native dogs, and only one lineage of rodents. Of some 209 native and regularly breeding species of birds, a little over half are found nowhere else! (Since most birds fly, this is an impressive percentage.) This island was also the home of giant predatory flightless birds—the elephant birds—as well as a number of larger lemurs, small hippos, and a giant eagle. Climate changes, beginning 2,500 years ago, and the arrival of humans, about 1,200 years ago, resulted in their extinction.

Geologists estimate that Madagascar first ripped loose from Africa about 160 million years ago, then separated from India about 80 million years later. Madagascar has major mountain ranges running along its entire eastern flank. Once blanketed by lush evergreen forests, these mountains produce a rain shadow to the west, where the vegetation is sea-sonally dry. The island's arid southwest supports an unusual spiny shru-bland, with odd succulents and thick-trunked baobab trees. Despite the proximity to Africa, the flora has ten families not found elsewhere. By latest count, Madagascar may have as many as 17,000 species of higher plants.[25] Of these, ferns number over 600 species, compared to only 500 species in all of Africa! Palms are well represented, with sixteen genera having 170 species; 165 of these are found nowhere else in the world. In fact, 96 percent of Madagascar's trees and shrubs are species found nowhere else.

Long isolation has also resulted in a very unusual fauna. With an area similar to California or France, Madagascar is home to 191 native terrestrial species of mammals, 346 species of reptiles, and over 300 species of frogs and toads.[26] While these numbers may not seem unusually high, *all the non-flying mammal species and nearly all the amphibian species are found nowhere else!* Clearly, Madagascar is distinguished by having one of the most unique biotas on the planet.

New Zealand is another island nation with a unique fauna and flora. Part of an active undersea ridge, formed where two tectonic plates abut, both the north and south islands have high mountains of recent vintage. Warm to cool-temperate, the islands are largely covered with evergreen forests and a flora whose relationships lie with Tasmania, Australia, and southernmost South America. Coniferous trees are especially well-represented in the forests of New Zealand. Indigenous flowering plants number about 2,400 species, with 86 percent endemic. Despite its isolation, New Zealand does not have a single endemic family of flowering plants; perhaps the flora developed too recently. New Zealand's fauna has a few outstanding endemics. Exterminated on the mainland, but still surviving on rocky islands off the coast, is the Tuatara (*Sphenodon*), an iguana-like reptile whose lineage has been distinct for almost two hundred million years. As in Madagascar, New Zealand was home to giant flightless birds, the moas. They vanished after humans arrived about eight hundred years ago. Today, a related chicken-sized flightless bird, the nocturnal kiwi, survives in dense forests.[27] All together, the native birds of New Zealand number 287 species, 74 of them endemic.

Isolation has resulted in the distinctive biotas of both New Zealand and Australia, but further on out, in the Pacific, we find the world's most isolated biota, that of **Hawaii**. Formed by a "hot volcanic vent" under the ocean floor, the Hawaiian archipelago has been affected by plate movement over more than forty million years. The most westward of the islands, Kauai, is also the oldest at approximately six million years. The youngest and most eastward of the islands, Hawaii, continues to grow with newly erupting lava. Because new islands formed over the "stationary volcanic hotspot," progressively younger islands are formed to the east as the plate moves westward. Thus, while today's oldest large island may be only six million years old, species could have spread from earlier islands—islands now eroded into the sea. Because of this historical progression and remote isolation, Hawaii is a unique laboratory for the study of dispersal, adaptation, and recent speciation.[28]

Isolated by more than 1,000 miles (1,609 km) from larger land areas, about 90 percent of Hawaii's thousand native plant species are endemic. Surprisingly, and despite the family's minute and easily dispersed seeds, Hawaii has only two native species of orchid! Similarly, and since they don't travel well across large ocean distances, there were no native freshwater fish, amphibians, reptiles, or mammals on the Hawaiian Islands! Unfortunately, isolation has resulted in vulnerability. New invaders, brought in by human activity, have led to the loss of many native species. Around sixty species of native Hawaiian plants survive with wild populations of fewer than a dozen individuals; they are not likely to be with us for long. Recent discoveries suggest that Hawaii was once the home of several smaller flightless bird species. Again, all vanished shortly after human colonization. Today, Hawaiian birds number 112 species, with 32 endemic and 54 recently introduced; the rest are oceanic travelers. Continuing introduction of plants, animals, and their diseases further threatens the native biota of Hawaii. No other part of the United States has been so vulnerable to invasive species.

Interestingly, after arriving on an island lush with vegetation and with few competitors, some groups expanded explosively. The fruit fly genus, *Drosophila*, has done exactly that: with approximately two thousand species around the world, almost one thousand are found only in Hawaii! This grand diversification, coupled with the fact that *Drosophila melanogaster* has been the subject of intense genetic analysis, has made Hawaiian fruit flies especially informative regarding both speciation and diversification. Zoogeographers place Hawaii and other isolated Pacific islands into a separate realm they call **Oceania**.

Chapter 5

PATTERNS, HOTSPOTS, AND THE GEOGRAPHY OF LINEAGES

Having surveyed species numbers around the world, let's continue by looking for unusual patterns of diversity. But before we
do that, we should re-examine the nature of the information available
in making our assessments. How do we actually quantify biodiversity?
Three measures have been given special attention regarding local biodiversity. We've already mentioned the first; the actual number of co-
occurring species in the area. This information is simple and is usually
available for the better-known lineages. However, there are problems
with even a simple species listing.

A primary concern is counting only those species which *properly
belong* to the fauna or flora. This requires being able to distinguish those
species that, while perhaps appearing indigenous today, were actually
schlepped in by human activities. Nowadays, escalation of travel and
commerce only adds to the difficulty of separating indigenous from non-
indigenous species.[1] Consequently, we will focus our attention on the
native species of a region—not recent arrivals.

Another measure of biodiversity is a bit more subtle; it is not the
number of species in an area but the number of *species found nowhere else*:
the **endemics**. And here we can run into taxonomic problems. Is a local
endemic species really that, or is it just a peculiar local variant of a more
widespread species? So-called "good species" are the result of careful
study, broad comparisons, and informed taxonomic decisions. Unfortunately, such a foundation is not available for many plants and animals.
Nevertheless, when a region claims a high percentage of endemics there
is the implication that we are picking up an important historical signal.

Finally, there is diversity at higher taxonomic ranks or more distantly related lineages. Imagine finding a small island with three species of mice on it and no other vertebrates. How would such an island compare with a similar island that had one mouse, one lizard, and one frog species? Surely, Mouse Island doesn't have the same level of biological diversity as does the island home of three different vertebrate lineages. From a conservation standpoint, there is an important difference. Mouse Island does not deserve as much effort as the island with three different vertebrate lineages. Think of these islands in terms of future potential; on Mouse Island, we can only hope for more mice.[2] But before we discuss diversity at higher ranks, let's return to simple species numbers and look at one of the most obvious patterns on the planet.

THE LATITUDINAL SPECIES-DIVERSITY GRADIENT

Starting at the equator, and moving towards either pole, overall species numbers decline. There may be areas where the trend is locally reversed——especially when traversing the great deserts——but the overall trend holds true. Species numbers are greatest in warmer regions, and progressively diminished toward the poles. This general pattern is found both on the land and in the sea.[3] Tree species numbers decrease steadily as we move northward toward the Arctic. Along the western Pacific we can expect to find forests with around 500 tree species in Taiwan, 250 species in southern Japan, about 80 in northern Japan, 20 on Russia's Kamchatka Peninsula, and only about 5 species above the Arctic Circle. Shifting to the western edge of North America, we can see a similar pattern in birds and mammals, only here we'll move southward. There are 222 species of breeding birds and 40 mammal species in Alaska. To the south, in British Columbia, Canada, we find 267 breeding bird species and 70 mammal species. In California, these numbers are 286 and 100. Moving to tropical Mexico the numbers explode, with 772 breeding birds and 491 mammal species. While some of this increase in numbers is due to Mexico's large area and diverse topography, there is a clear tropical effect. Butterfly numbers are just as dramatic. All of North America north of central Mexico is home to about 750 butterfly species; to the south, the Neotropical realm numbers an estimated 7,500 species; a ten-fold difference!

Of course, the latitudinal species diversity gradient has exceptions; some groups of plants and animals are more diverse in the cooler temperate zone. The rose family, the pines and birches, oaks, aphids and salamanders, all have more species in temperate areas than in the tropics. But these are exceptions. The generalization holds true: as you leave the tropics and move toward the poles, species numbers decline. This is especially visible in some iconic tropical plants. Palms, with over 2,500 species, have a very limited frost tolerance; coastal North Carolina is as far north as they naturally grow in North America. Even more intolerant of cold are the gingers, the bananas, and their kin. Often with huge leaves and colorful inflorescences, these plants give moist tropical vegetation much of its exotic appearance. But these are all specific examples; what might be the general causal factors for increased tropical species richness?

Surely simple biological productivity has to be a key factor, whether measured as annual carbon fixation or standing biomass. Given sufficient moisture and agreeable temperatures, more direct sunshine means more photosynthesis (more carbon fixation) and that means more fuel to keep the ecosystem running. In addition, neither cold nor drought is conducive to life activities. Plants can't photosynthesize while they are freezing or when they are parched. A clear example of the tropical advantage is seen in leaves. There is nothing in cooler northern climes that compares with the leaf of a banana plant or coconut palm. Large leaves are expensive to produce, and you find them only in the evergreen tropics, where they can continue photosynthesizing over several years.

This brings us to the problem of **evapotranspiration,** a measure of moisture loss from plants. When things get too dry, pores in the leaves close and photosynthesis shuts down. In severe cold, the plant cannot pump water up from its roots and, again, the plant must go into dormancy. Thus, **seasonality** is a critical determinant of plant productivity. Even though they may feed on ocean fish, a study of island-living birds around the world concluded that **climate**——temperature and precipitation——is the most important determinant of species numbers.[4] Staying with birds, frigid Greenland is home to 57 species of breeding birds, while the state of Georgia in the eastern United States numbers 160 species. Further south, breeding birds find their zenith in Colombia, with around 1,600 species. Here's as dramatic an example of the latitudinal species-richness gradient as you are likely to find.

The geographical range of species is another factor contributing to the latitudinal diversity gradient. For example, if you are a species that can survive a typical Midwestern year, from subfreezing winter through torrid summer, you are likely to be able to range over a wide area. But if you are adapted to the cool misty interior of a tropical montane cloud forest, you are unlikely to be successful in the hot lowlands, or in any other stressful climate. In addition, it is unlikely that your seeds can reach a similar habitat on a distant mountainside. For this reason, montane cloud forest species often have very limited geographic ranges. Ecotourists visiting Costa Rica's Monteverde Cloud Forest Reserve can experience this unusual climate for themselves. Tourist hotels near the reserve offer comfortable first-class accommodations, and yet they do this without heating or cooling systems! At 1,500 m (5,000 ft.) above sea level, Monteverde provides a truly temperate experience. It's warm every afternoon and cool every evening, throughout the year!

Beyond the cloud forest and more generally, tropical environments support more "specialist" species than do strongly seasonal environments. In a recent comparison of butterflies, tropical forest caterpillars were shown to have more specialized diets than their temperate forest counterparts.[5] This means more species can coexist without getting in each other's way. A more dramatic example is found among Costa Rica's mammals; a large number of species are fruit-eating and nectar-slurping bats! Such lifestyles are only possible in an environment where flowers and fruit are available throughout the year. Having worked with Costa

Rica's flora over more than three decades, I was impressed by how much that flora differs from the floras of nearby Nicaragua or Panama. That's not what you see when you leave Indiana to visit Illinois to the west, or Ohio to the east. We had thought that Illinois had a single endemic flowering plant: the Kankakee mallow (*Illiamna remota*). But then this same species was found in West Virginia, and Illinois lost its only living endemic flowering plant. Midwestern floras are nearly identical, and why shouldn't they be? Large areas of the Midwest were sitting under an ice-sheet 20,000 years ago. Nearly all these plants are recent re-colonizers. Nicaragua, Costa Rica, and Panama have suffered no such indignity; their species have been free to sit where they are and diversify over millions of years. All three countries have many endemic species. Simply put, a greater percentage of tropical species have more limited geographic ranges, compared with temperate or polar species. Called Rapoport's rule, this generalization applies to both plants and animals.[6]

A recent paper in *Science* looked at the latitudinal diversity gradient from a genetic perspective. These authors compared closely related species pairs of birds and mammals in both tropical and temperate environments, and estimated their times of divergence by using DNA sequences.[7] The result was surprising and significant: tropical species pairs are generally further apart in time than are bird and mammal pairs from higher latitudes! Speciation is *more rapid in cold climates* than in tropical latitudes! At first, this seems counterintuitive. However, if species are going extinct more rapidly in cold climates, more vacant niches will be available for new species to invade. A higher **extinction rate**, followed by speciation, means that local turnover rates are higher at high latitudes. Certainly, extinction is a significant factor underlying the latitudinal species-diversity gradient. Just imagine what conditions were like in the American Midwest or central Europe during the height of the last glaciation.[8]

And then there's something else we don't like to talk about: parasites and pathogens. Our own species is a fine example. If you would like to get really sick, let me suggest visiting moist lowland tropical Africa. There are two reasons why Africa has so many human maladies. The first is the climate: many of our pathogens and parasites can't survive a northern winter. The second reason is historical: Africa is where our species originated and where our closest relatives reside. Here, our diseases had ample time to make us their hosts. Here is where the HIV virus originated, where malaria, yellow fever, liver flukes, and many other human pathogens abound. This same problem affects other species, both animal and plant: there are many more pathogens, parasites, and herbivores in the tropics. In summation, the latitudinal species-diversity gradient seems to be the simple product of more benign environmental factors and more competitive challenges in tropical latitudes.

AREA AND ALTITUDE EFFECTS ON BIODIVERSITY

The latitudinal species-diversity gradient is the most prominent area-diversity relationship on the planet. Also, and in general, as one moves from smaller to larger land surfaces one finds more and more species. This is the species-area relationship, and it usually forms a simple linear gradient from smaller to medium-sized areas. As one reaches ever-larger areas, however, fewer new species are encountered, and the graphed relationship tends to flatten out. As you might expect, the latitudinal species-diversity gradient interacts with the species-area diversity gradient. A recent study of North American vascular plants found that species-area relationships change with latitude. That is, as you go further north, species tend to have wider ranges; some are even circumpolar.[9] An Illinois prairie can have as many as 300 species of higher plants in the prairie and its moist depressions. Then, as you enlarge your survey westward to include Iowa, you do pick up new and different species. But you add only a small number. As we mentioned, this is one way in which the tropics really are different. Beginning in Costa Rica starts you off with lots of species, but then, as you expand to either Nicaragua or to Panama, you will be adding many species you hadn't encountered in Costa Rica. Here, adding the local and wider areas really does escalate the numbers.

A striking example of the area-diversity relationship comes from a study of islands in the Caribbean, from small to large. With both species numbers and island area graphed logarithmically (1, 10, 100, etc.), the relationship results in a straight line, showing clearly how greater area supports greater diversity.

TABLE 5-1: SPECIES AREA RELATIONSHIPS AMONG PLANTS AND TREES

Geographical Unit	Area in km^2	Plant Species	Tree spp/ha
Reunion Island	2,510	660	40
New Caledonia	16,700	3,061	97
Madagascar	587,000	11,000	146
New Guinea	808,000	15,000	228
South America	18,000,000	90,000	283

From a talk by E. G. Leigh Jr. at Field Museum, April 2007

An interesting aspect of the area-diversity relationship is that, on a small scale, the tundra can outnumber the rain forest. If you compare a square meter of tundra to a square meter on the floor of the rain forest, you'll find more species in your tundra quadrant! There will be mosses, liverworts, lichens, sedges, and surely a grass or two in your tundra plot. The dark floor of the rain forest may have a few seedlings, a sapling, and lots of rotting leaves, but not much more. In the rain forest, of course, much of the diversity sits high in the canopy. Even so, one reason the square meter of tundra is so rich is that the plants are so small. And the reason they are so small is because the environment is so severe. On a smaller scale (one to a hundred square meters) arctic and temperate vegetation can be impressive in their species numbers. However, it is at larger scales, where the moist tropics reign supreme—with bigger plants carrying lesser plants upon them. In fact it is the accumulation of little plants—growing as epiphytes on trunks and branches—that raises the numbers of plant species significantly in evergreen tropical wet forests. In an altitudinal study, ranging from 30 to 2,600 m elevation on the Caribbean side of Volcán Barva in Costa Rica, 264 species of ferns were found. Of these, 96 were terrestrial species, while 121 were recorded as low-trunk epiphytes, and 113 were found as high-canopy epiphytes. Though some species were recorded in more than one category, the epiphytic ferns greatly outnumbered the terrestrial ferns.[10] Another recent survey estimated that around a quarter of all the species of native flowering plants in Costa Rica are epiphytes, and almost a half of Costa Rica's endemics are epiphytes![11] In fact, Al Gentry and Cal Dodson suggested that the extraordinary floristic diversity of the Neotropics was largely due to its epiphytes.[12]

A terrific way to experience the species-diversity gradient is to ascend from lower to higher elevations in the tropics. Again in Costa Rica, and studying ferns, Jürgen Kluge and Michael Kessler enumerated species along an elevational transect ranging over almost 3,400 m (11,150 ft.).[13] They found about 20 species per plot at 100 m elevation, around 50 fern species in plots between 1,200 and 2,500 m, and around 10 species at 3,000 m. Both total fern species numbers and endemics were highest between 1,000 and 2,500 m elevation! Researchers in Borneo found a similar diversity pattern for moth species; the highest numbers were found between 500 and 1,000 m elevation.[14] As in many other tropical studies, there is a clear "hump in species richness" at middle elevations. Because temperatures fall by about 6°C for every 1000 m (3°F per 1000 ft.) above sea level, air forced up the mountain slopes expands and becomes cooler. This cooler air cannot hold as much moisture, resulting in frequent rain and misting over much of the year. It is in these montane "cloud forests" where trees and their branches are festooned with smaller epiphytes—perhaps the richest biome on the planet.

It should be no surprise that the highest elevations of tropical mountains have low species numbers. Smaller areas and colder, more windy conditions are the reason. English speakers developed a phrase for the high elevation habitats they experienced in East Africa: "Summer every day and winter every night!" Above about 3,000 m (10,000 ft.) elevation, a clear night sky can drop temperatures to freezing. In this way, high tropical mountains offer a greater variety of temperature and moisture regimes than do the more uniform conditions in the hot lowlands. Varied altitudinal zones, in turn, allow for greater speciation, as ecological selection promotes diversification. Birds of Paradise living in New Guinea's mountains maintain high species numbers by living at different elevations, in what is called *altitudinal stratification*. By affording a great variety of habitats, mountain systems are a major reason why our planet is so rich in biodiversity.

NUMBERS OF SPECIES AND NUMBERS OF INDIVIDUALS

In both tropical and subtropical habitats, superabundant and widespread species tend to be few, while the great majority of species—animal or vegetable—are rare. This is especially evident in some of the large tropical trees, where you have to hike some distance to find another tree of the same species. And it is not just trees; many shrubs, herbs, and animals seem "always to be rare." The meter-tall shrublet *Piper veraguensis* is very distinctive: its foot-long leaves hang vertically from leaf-stalks attached near the center of the blade, where smaller veins form circles around the stalk attachment. Though known from Costa Rica to Colombia, I have seen this species only twice in many years of field work in Costa Rica. Exactly how a species can manage to persist at such low population densities over so wide a range is a mystery to me.

Eric Dinerstein describes the importance of rarely encountered species, both for biodiversity and conservation, in his book *The Kingdom of Rarities*. He points out that some species are "doubly rare"—that is, both geographically restricted and rarely encountered in those areas where they do live. Another pattern is to travel for some distance through forests and over mountains and not see a particular species, only to come upon a particular valley and find it in considerable numbers. We botanists describe such a species as being *locally abundant*, while zoologists tend to use the word *patchy*. In an extensive study of Amazonian trees, a recent analysis found that only 227 species were "hyperdominant," accounting for half of all the trees, while 11,000 species were rare.[15] Surely, rare species are an important way of squeezing more species into the same geographical region. While these patterns have to do with contemporary species-packing, there are also distribution patterns with deep historical roots.

PLATE TECTONICS AND PATTERNS IN BIODIVERSITY

Scientists, like other human beings, are social creatures. If everybody in your discipline proclaims that the world's continents couldn't possibly move, you are not likely to sing a different song. Geologists in the United States held fast to their belief in the *stability of the continents*. But Alfred Wegener was a German meteorologist, and he was impressed by the shape of the Atlantic Ocean, with almost symmetrical eastern and western shores. Also, he had read the work of paleobotanists, who spoke of an ancient floristic Gondwanaland. Rocks from Gondwana (a little town in southern India) contained fossil leaf impressions that included a variety of different trees. One species, *Glossopteris*, had leaves rather like those of a willow. But there were also fern-like leaves and several other distinctive species as well. This 220 million-year-old fossil assemblage became known as the Glossopteris flora, characterized by a consistent suite of distinctive species. The co-occurrence of these tree species implied that they lived in the same kind of forest. And that was the problem! Rocks of similar age with this same suite of plant fossils were also found in Australia, South Africa, and South America!

Surely it's possible that one or two species might have been growing around the world in those ancient times——but all these species, always together? That seemed highly unlikely. Paleobotanists concluded that the Glosopteris flora had been a *single forest spread over a single continent*, calling this ancient continent Gondwanaland or **Gondwana**. With these and many other data, Alfred Wegener fashioned his theory of "continental drift" in 1915.[16] *Impossible!* was the response of geological leaders in

North America. After all, how could continents rupture or move around? Good scientists do good science, and continents drifting around the surface of the Earth was, clearly, bad science![17]

But then, careful surveys of the ocean depths in the 1950s and 60s produced unforeseen discoveries. Nowhere did the ocean floor appear to be more than 250 million years old. This was totally unexpected! Furthermore, the floor of the deep oceans was crossed by long volcanic fissures. The Atlantic Ocean has a volcanic ridge-system *running exactly down its center*, all the way from Iceland to the Falklands! At about this same time, it had become known that Earth's magnetic field can flip, with the north magnetic pole becoming south, and back again. We still do not understand how or why the magnetic field changes in this way, but the reversal of magnetic poles could now be used to help date volcanic rocks! As lava cools and hardens, its magnetic polarity becomes frozen in place, and we can tell its north/south polarity at that place and time. Using this data to analyze lava beds east and west of undersea fissures produced an astounding result. Volcanic rocks on opposite sides of undersea ridges proved to be *perfectly symmetrical* in their pattern of magnetic reversals! The reversals on one side were the mirror image of those on the other side! Suddenly, geologists realized hot magma was welling-up along the centers of these undersea ridges, and then spreading laterally on both sides—producing symmetrical magnetic bands of lava rocks on either side as they moved apart. The mid-Atlantic ridge, by producing new magma at its center, has been expanding the ocean floor laterally, east and west!

Coupling the near-symmetrical continental shelves of the Atlantic with the mid-Atlantic ridge's continuing east-west expansion made something very clear: over millions of years, the eastern and western margins of the Atlantic Ocean were being continuously pushed further apart. Not all that long ago in geological time, *there had been no Atlantic Ocean*. Earlier, there had been a huge southern continent—**Pangaea**—on which the Glossopteris flora flourished. That land mass and its vegetation had been torn apart, just as Wegener imagined. Our planet's surface was made up of shifting plates! Surely one of the grandest revolutions in the history of modern science, the theory of **Plate Tectonics** now gives us deep insight into how the continents have changed over time. Little Gondwana and the Indian subcontinent on which it sits had been ripped loose from a larger southern land mass, then "rafted across" the Indian Ocean, only to smash into the Asian continent, driving the Himalayan mountains high into the sky.[18] The theory of plate tectonics is Wegener's continental drift confirmed!

The notion of continents being torn apart and carrying their biotas with them became a new paradigm among certain biogeographers. Ridiculing the notion that plants and animals might disperse across vast oceans, these researchers promoted "vicariance biogeography." Why do the flightless ratite birds all live on southern continents? Because their original home was in Gondwana, claimed these biogeographers. Unfortunately, recent DNA comparisons indicate the ancestors of these birds *flew* to their present homes.[19] Lung fishes tell a more compelling story, with survivors living only in Australia, South America, and Africa. These do seem to be sitting on the broken parts of their original landmass. Vicariance biogeographers disparaged suggestions of cross-oceanic dispersal as untestable *ad hoc* hypotheses. Things have calmed down since those acrimonious debates in the 1980s, and people are now comfortable with the fact that some lineages have managed to travel across the oceans on their own, while others have held tight to their continental moorings.[20] In fact, the idea of continental movements "having carried their living passengers with them" explains the world's most unusual animal disjunction.

WALLACE'S LINE

Supporting himself by collecting animal specimens, Alfred Russel Wallace traveled extensively between Malaya, Indonesia, and New Guinea. While collecting, he noticed something very unusual. Over a distance of as little as 48 km (30 miles)—between the Indonesian islands of Bali to the west and Lombok to the east—he had collected very different animals. There was a similar distinction in animal distributions between the islands of Borneo on the west and the Celebes to the east. Monkeys, cats, deer, rhinoceros, and squirrels could be found on the western side of this divide. To the east, Wallace found a world of strange arboreal marsupials, many lizards, birds of paradise and their relations—but no monkeys, no cats, no deer. Though a number of animal lineages cross **Wallace's Line,** the larger picture holds true. Here is a major division in animal distribution—a disjunction—over a very short distance!

How could two faunas be so close, and yet so distinct? Enter plate tectonics! As in the case of India, the Australian plate, carrying New Guinea and Australia with it, has moved northward from more southern climes. While not smashing into its northern neighbors with the force and effect of the Indian subcontinent, plate movement did bring a unique southern biota into close proximity with a northern fauna. This scenario explains why birds of paradise and bower birds are abundant in New Guinea: there are no cats to eat them. Thus, an explanation for two very different faunas in such close proximity is that two long-separated tropical faunas were brought into close proximity by tectonic movement.

Unlike the animal disjunctions, however, Wallace's Line is difficult to discern when examining the vegetation. Plants appear to have sent their disseminules back and forth across these marine barriers with great ease. Only a few nonflying terrestrial mammal families are shared between Southeast Asia and Australia. In comparison, sixty-nine flowering plant families are shared over these same areas. Plant species don't tell the same story as the animals.

Here is an instance where higher taxonomic levels, with a longer history, do retain a historical signal. Tropical American ferns number about 1,270 species, not very different from the 1,235 species listed for Southeast Asia, Australia, and the nearby islands (Australasia). However, the Australasian flora has 93 families of ferns and fern allies, while the Neotropical flora has only 38 families. That's a huge discrepancy! Australasia has more than twice as many families as all of tropical America. Fern families, a much higher rank than species, are telling us something. Twice as many fern families in Australasia supports the idea that two long-separated floras have come together and joined in this part of the world. Widely dispersed by minute airborne spores, species of the original Southeast Asian and New Guinea/Australian fern floras are no longer discernable.

THE EASTERN ASIA–EASTERN NORTH AMERICA DISJUNCTION

Asa Gray, professor of Botany at Harvard University in the mid-1800s, was the first to notice something unusual in northern plant distributions. Shipments of plant specimens from Japan and China often included genera also growing in the eastern United States. As collecting continued, something became very evident: there was a clear similarity in the floras of eastern Asia and eastern North America. In fact, for many groups, there was greater similarity between eastern North America and eastern Asia than between eastern North America and western North America! The flora of Oregon lacks many of the iconic plant genera shared by eastern Asia and the eastern United States. Tall and graceful, only two species of tulip trees are found in the world today: *Liriodendron tulipifera* in the eastern United States and *Liriodendron chinense* in eastern China. Our sassafras trees, often with mitten-shaped leaves, are members of a genus with three species: two live in eastern Asia and one is in eastern North America. Skunk cabbage (*Symplocarpus foetidus*) of our eastern swamplands is found in only two areas of the world: eastern Asia and northeastern North America. The same pattern holds true for lopseed (*Phyrma leptostachya*), a forest-floor herb.[21]

Shifting to animals, the strange paddle fish of the Mississippi River is a member of an ancient lineage with only one living relative, swimming in China's Yangtze River. A foot-long salamander that looks like it's been run over by a truck, the hellbender, lives in the Ozarks and Ohio River drainage; its only close relative lives in eastern Asia. The world has only two species of alligator—one in the southeastern United States and the other in southern China. What might explain these concordant disjunctions?

Twenty-million-year-old fossils tell us that Europe did have a much richer biota in those earlier times. The dawn redwood (*Metasequia*) was known as a fossil in northern regions around the world but was found as living populations in China only later. A rich biota of tulip trees, alligators and odd amphibians were all part of a northern "mixed-mesophytic forest" before the ice ages began. Since then, glaciation severely reduced the flora of northern Europe and, as we already noted, left eastern Asia the least diminished. Surviving species in eastern Asia and eastern North America have given us the biogeographic disjunction first noticed by Gray.[22] Here again, history has left a biogeographic signature.

THE CONCEPT OF "BIODIVERSITY HOTSPOTS"

From a larger perspective, there are regions in the world where biological diversity is unusually high. A few of these areas have been labelled "hotspots" for the purposes of biological conservation. Unfortunately, and rather like beauty, special areas of biodiversity may lie in the eye of whoever is making up the listings. Criteria for assessing special regions are varied. How significant are overall species numbers in such determinations, and which plant and animal groups will be measured? Should species unique to the region—the endemic species—play a significant role in designating a hotspot? Should threat of imminent extermination be an important criterion? Though arbitrary, the notion of **hotspots** has been grandly envisioned in a large book with many stunning photographs: *Hotspots: Earth's Biologically Richest and Most Endangered Terrestrial Ecoregions*, by Russell A. Mittermeier, Norman Myers, and Cristina Goethsch Mittermeier.[23] Using environmental threat and endemic species, these authors listed their hotspots. Inevitably, some choices seem arbitrary; New Caledonia and New Zealand are included, as are the islands of the Caribbean, but not New Guinea, where species numbers are high but threats are not as imminent. Also the areas covered by each hotspot vary greatly. Much of the central and western Pacific is assigned to the Polynesia/Micronesia hotspot—a huge area—but little New Caledonia is a hotspot all its own. Such choices are inevitable in any survey designating only twenty-five hotspots (Table 5-2).

TABLE 5-2: BIODIVERSITY HOTSPOTS
(MITTERMEIER ET AL., 1999)

Mediterranean floras are marked by an asterisk.

Region	Vascular Plant Species	(endemics)	Birds	Mammals
Caribbean	12,000	(7,000)	668	164
Mesoamerican	24,000	(5,000)	1,193	521
Choco-Western Ecuador	9,000	(2,250)	830	235
Tropical Andes	45,000	(20,000)	1,666	414
Brazil, Atlantic forest	20,000	(6,000)	620	261
Brazil, Cerrado forests	10,000	(4,000)	837	161
Chile, central region*	3,429	(1,605)	198	56
California, chaparral*	4,426	(2,325)	341	145
Africa, Eastern Arc Mtns.	4,000	(1,400)	585	183
Africa, Cape flor. prof.*	8,200	(5,682)	288	127
Africa, succulent Karoo	4,849	(1,940)	269	78
Africa, Guinean forests	9,000	(2,250)	514	551
Madagascar	12,000	(9,704)	359	112
Mediterranean Basin*	25,000	(13,000)	345	184
China, South-Centr. Mtns.	12,000	(3,500)	686	300
Asia, Caucasus Mountains	6,300	(1,600)	389	152
Asia, Indo-Burma area	13,500	(7,000)	1,170	329
Asia, W. Ghats and Sri Lanka	4,780	(2,180)	528	140
Asia, Philippines	7,620	(5,832)	556	201
Asia, Sundaland	25,000	(15,000)	815	328
Asia, Wallacea	10,000	(1,500)	697	201
Australia, Southwest*	5,469	(4,331)	181	54
New Zealand	2,300	(1,865)	149	3
New Caledonia	3,332	(2,551)	116	9
Polynesia and Micronesia	6,557	(3,334)	254	16

Numbers of vascular plants, mammals, reptiles, birds, and amphibians, are the basis on which these hotspots were chosen; these organisms are the ones we know the best, and the ones in which endemism is easy to assess. We've referred to these "hotspots" earlier and, as noted, one of the most significant criteria followed by Mittermeier, Myers, and Mittermeier for designating hotspots is advancing agriculture and continuing deforestation. In these circumstances, narrow-ranging endemic species are often the most susceptible to extirpation. The Caucasus Mountains, Africa's Eastern Arc Mountains, Madagascar, and the Philippines were all designated hotspots because of the threats they face. With the purpose of drawing world attention to the grave dangers being faced by specific areas, hotspots continue to play a role in conservation priorities. However, Peter Kareiva and Michelle Marvier have argued that our world also harbors important "biodiversity coldspots." Yellowstone National Park may not have a large number of species, but it is the only place in the coterminous United States harboring the last assemblage of once widespread large mammals, such as grizzly bears, elk, mule deer, wolves, puma, and bison.[24] Clearly, we need to preserve as many species as we can, and in every corner of the world that we can. That said, let's try and understand why some areas manage to support unusually high species numbers. We'll begin with little Costa Rica, which has played an important role in both tropical biological research and conservation.

COSTA RICA: A LITTLE COUNTRY WITH LOTS OF SPECIES

Costa Rica is a splendid example of tropical biodiversity. The people of Costa Rica have supported a vigorous intellectual tradition throughout their history. After World War II, they abandoned their army to spend more money on education. After many decades, and thanks to both resident biologists and visiting researchers, we now know the flora and fauna of Costa Rica quite well. Located in Central America's narrowed southern isthmus, between Nicaragua and Panama, this country is completely tropical. Bounded by two large oceans, and averaging only around 160 km (100 miles) in width, Costa Rica receives ample rainfall over its entire area. As mentioned earlier, Costa Rica is home to about 10,600 species of vascular plants. If you take that number and divide it by the number of square kilometers ascribed to Costa Rica, you get a larger figure than you do for any other country on the planet! Colombia, Brazil, and China have many more species, but they support them over much larger territories. No other country in the world packs so many species into so small an area as does Costa Rica.[25]

Though half the area of Ohio—on a flat map—Costa Rica boasts twice as many species of mammals as all of North America north of Mexico. Likewise, this area of about 51,100 km² (19,730 sq. mi.) has twice as many species of ferns as does all of North America north of Mexico. Costa Rica's bird count stands at 878 species, compared to about 850 in all of northern North America. We can amplify these species numbers with 175 amphibians and 218 reptiles, to go along with 228 mammal species. With such numbers and a very cordial populace, it's easy to see why this little democracy has become so popular a destination for ecotourists.[26]

What is it about little Costa Rica that allows so many species of plants and animals to make it their home? Simply stated: a tropical setting with generous rainfall and high mountains. The driest part of the country, the northwestern province of Guanacaste, averages about one meter (39 inches) of rainfall each year. Falling from May to November, these rains support a deciduous woodland at lower elevations. Here, few trees retain their leaves over the hot, five-month-long dry season. At higher elevations (above about 800 m / 2,625 ft.) temperatures are cooler, evapotranspiration is diminished, and partly evergreen forests are the original land cover. With higher rainfall and shorter dry seasons, evergreen broad-leaved forests originally covered all the rest of Costa Rica, except at the highest elevations. On the Caribbean side of the country, rain falls virtually all year long, supporting swamp forests as well as lowland rain forest. However, these many lowland forests are only one factor accounting for Costa Rica's high species numbers.

Costa Rica is a *mountainous* country. Its volcanoes and elevated highlands are part of "the ring of fire" bordering much of the Pacific Ocean. With the Caribbean plate overriding the Pacific margin, mountains are still rising and earthquakes are frequent. Beginning in the west, near Nicaragua, with a few smaller volcanoes, Costa Rica has a central chain of mountains traversing its entire 480 km (300 mile) length. The Cordilleras of Guanacaste and Tilarán dominate the northwestern third of the country, with both active and older volcanoes. A central volcanic group forms the nation's midsection, dominated again by volcanoes old and new. Finally, in the eastern half of the country, we have a massive uplift— the Cordillera de Talamanca—merging with the Chiriqui highlands of western Panama. Costa Rica's highest elevation, Cerro Chirripo, reaches 3,420 m (11,220 ft.) above sea level. All these cordilleras support tropical rain forest along their lower foothills, cooler cloud forests on their mid-elevation slopes, montane forests above about 2,200 m (7,280 ft.) elevation, elfin woodlands along exposed windy ridges, and grassy alpine *paramo* formations above about 3,200 m (10,500 ft.). These varied highlands, and the many valleys descending their slopes, provide the scaffolding for little Costa Rica's spectacular biological diversity. As we've noted, cool, moist montane cloud forests are especially rich in epiphytes. These are mostly smaller plants growing on the trunks and branches of larger trees; little animals also live in these tree-top communities.

Mountains make a difference. Of Costa Rica's 878 bird species, 160 are found only at higher elevations. As we noted in the fern studies, many plants and animals are restricted to the cooler highlands, and some of these are found nowhere else in the world. Of Costa Rica's birds, forty-seven are unique to the country. Many other species, of course, are wider ranging, including migratory birds that spend our cold winters in tropical climes. Over time, Central America's narrow isthmus has served as a bridge between two grand biotas, having been enriched from both the north and the south. On Costa Rica's highest mountains, we find plants characteristic of northern climates and also the high Andes; these are plants that cannot compete in warmer, lower elevation vegetation. Oaks dominate the cooler high-montane forests; they are a northern element in this southern flora. (Oaks are diverse in the United States and Mexico, diminishing in numbers in southern Central America, with only one species reaching northern South America.)

To summarize, high mountains with varied and dissected topography, together with a tropical setting receiving ample rainfall with differing local seasonality, help Costa Rica support so many species. The separation of related species—both plant and animal—by "altitudinal stratification" is an especially important factor supporting diversity in this little republic. These same factors escalate as we move south into the Andes.

THE ANDES MOUNTAINS OF SOUTH AMERICA

South America is also home to many "hotspots" of species richness. The eastern slopes of the Andes, bordering the Amazon and Orinoco drainage basins, are a huge area, with conditions very similar to those of Costa Rica. The nations of Colombia, Ecuador, Peru, and Bolivia are home to very high numbers of plants and animals. Here, a warm climate is coupled with frequent rain along the eastern slopes of the world's largest tropical mountain system. Colombia has the highest recorded bird count in the world, with around 1,800 species. Supporting Colombia's biodiversity is: (1) the exceptionally rainy *Choco* region of the northwest; (2) the high Andes, split into three parallel cordilleras; (3) open grasslands in the Orinoco basin; and (4) the Amazonian flora in the southeast.

As in Costa Rica, higher elevations along the eastern flank of the Andes support cooler evergreen forests often shrouded in mist and fog. Prevailing winds force warm moist air upward along the mountainside; becoming cooler as it rises, this air can no longer hold as much moisture, resulting in frequent rain, drizzle, and misting. With humidity high for most all the year, branches and tree trunks become festooned with epiphytic mosses, leafy liverworts, ferns, orchids, aroids, and bromeliads. "Tank bromeliads" are especially significant in enhancing biodiversity. Their broad clasping leaf-bases hold little pools of water high in the canopy. Here, far above the forest floor, animals can find a drink of water, and smaller aquatic creatures can live out their entire lives. Edward Wilson declared epiphytes "biodiversity multipliers." Not only do these plants add their own numbers, but by providing special homes for little animals they further elevate species richness.[27] That's the good news; the bad news is that the weight of epiphytes often causes branches to break, resulting in trees considerably shorter than those in lowland rain forests. Nevertheless, cloud forest may be the richest biome of all, supporting more species per hectare than even lowland tropical rainforest.

We mentioned the high species numbers in Colombia, Ecuador, and Peru, but the complex Andean highlands have been well collected only along major roads and near towns. Also, much of our knowledge of this region comes from isolated nature reserves and the research stations they support. One of the largest nature reserves along the eastern flank of the Andes is Manu National Park in southeastern Peru. Overall, the park covers about 1,532,806 km² and ranges from 300 to 4,000 meters in elevation. Rainfall is sufficient to support evergreen forests throughout all but the highest elevations; the dry season lasts about three months. Overall, higher plants probably number around 6,000 species.[28] These numbers include the tropical lowlands, cooler montane forests, and high paramo formations. The number of bird species recorded over this same altitudinal range in 2006 was 1,005, the highest number for any preserve of similar area on Earth. The number of mammal species recorded at Manu is 222, including 92 bat species, 58 rodents, 21 carnivores, and 14 primates. Significantly, this survey found not a single new species of birds, but it did include at least twelve new mammal species.[29] This is a clear example of how much better we know the birds than we know the more secretive mammals. And that's not all: nineteen species of murid rodents were found to be carrying fifteen species of lice, three of which had never been seen before![30]

Because of the difficulties in traversing Manu's full altitudinal range, most surveys have been limited to lower elevations along the river near the Pakitza research station. Here, 1,300 species of butterflies have been captured and identified (as of 1995).[31] Similar studies found 498 species of spiders in 33 families and 136 species of dragonflies and damselflies. Amphibians and reptiles numbered 128 species, including one crocodilian and 67 frog species. All these numbers were based on short-term studies. Longer term studies at another site in Peru resulted in finding

113 species of amphibians and 118 species of reptiles, indicating that there are many more species to be found at Manu.[32] All told, the moist eastern slopes of the tropical Andes probably support the richest repositories of biological diversity in the world.

OTHER MOIST-TROPICAL HOTSPOTS AROUND THE GLOBE

As we've noted, locations with high concentrations of species diversity can be characterized by only a few fundamental variables. These factors are "high energy input with constant water supply and extraordinarily high spatiotopographic complexity."[33] Little Costa Rica and the eastern slopes of the Andes have this combination of tropical warmth, ample rainfall, and complex mountainous terrain. Other parts of the world are similarly blessed, especially in Southeast Asia.

The island of Borneo (Sabah) is famous for being home to unusually high numbers of plants and animals. Kinabalu Park supports between 5,000 and 6,000 species of vascular plants. This same small area of 737 km² is home to 326 species of birds and more than 700 species of orchids. In these forests one can find many species of vining insect-digesting pitcher plants. Other nearby islands, such as Indonesia, the Celebes, and New Guinea, are also extremely rich in species numbers and in endemic species. Of the twenty-five hotspots designated by Myers and Mittermeier, four are located in southeastern Asia. The large Indo-Burma hotspot ranges from the southern slopes of the Himalayas and easternmost India, through Myanmar to Vietnam.[34] The Sundaland hotspot ranges from the southern half of the Malay peninsula to the islands of Java and Borneo.[35] The Wallacea hotspot includes the eastern islands of Indonesia, the Celebes, and islands west of New Guinea.[36] The Philippines hotspot is restricted to the islands of this archipelago.[37]

Important tropical wet-montane biotas are also found in Mt. Cameroon of western Africa, and on the higher mountains of East Africa's rift system, especially along the Uganda-Congo border where the mountain gorilla survives. In Asia, moist highland forests are found along the southern flanks of the Himalayas, from Nepal through Burma, Thailand, Laos, Vietnam, and southernmost China. Species-rich montane hotspots are also found in the Western Ghats of India and in Sri Lanka. While New Guinea supports one of world's most unusual faunas, it is less threatened by human expansion.

SPECIES RICHNESS IN SEASONALLY DRY OR COLD FORESTS

Deciduous broadleaf forests of both the temperate and tropical zones also contribute to our planet's enormous species count. Despite severe winters, temperate broadleaf forests are important regions of biodiversity. We've already noted that the Caucasus Mountains have been designated a hotspot, and China's southwest mountains support high species richness. Nevertheless, some of the world's most impressive temperate forests are not especially rich in species. The Giant Redwood forests of northwestern California, and the coniferous forests of the Pacific Northwest, dominated by Douglas fir and western hemlock, are awesome in their height and impressive in the density of tree trunks. No other forests in the world have so many tall slender trunks packed so closely together. Yet these grand forests are poor in species. No matter where they are found, forests dominated by the cone-bearing Gymnosperms do not support high species numbers. Perhaps this is a consequence of the chemicals with which they protect themselves. Conifer leaves are less subject to insect attack than are the leaves of flowering plants; fallen conifer logs rot much more slowly than do their flowering plant counterparts. Similarly, the great needle-leaf forests of the taiga are poor in species numbers, but here a long and cruel winter plays an important role. Moreover, just as the seasonally frigid temperate forests contribute to the world's overall biodiversity, seasonally dry tropical regions also enrich the world's species numbers.

Seasonally dry tropical woodlands and thorn bush are often rich in endemic species, and many have been designated hotspots on that account. While species numbers may not be as high as in the wet tropics, many local endemics live in these seasonally parched landscapes. The floras of the Sonoran Desert in the southwestern United States and northwestern Mexico, as well as the unusual flora of the nearby Baja California Peninsula are quite rich in species. In South America, the *chaco* vegetation of northeastern Brazil, and the *cerrado* and *caatinga* formations of Bolivia and southern Brazil are all rich in endemic species. The *cerrado* includes an estimated 10,000 vascular plant species, with many having become adapted to intense fires over the last five million years.[38] In Africa, the *Acacia-Commiphora* thorn bush in eastern Ethiopia and Somalia, and the semi-desert floras of the Karoo and Namibian deserts are outstanding in harboring unusual plants and animals. The vegetation in these areas varies from open woodlands of short trees to scattered thorn scrub on rocky soils, with grassland covering flat, fire-prone sites.

We've discussed an unusually rich biome in subtropical, seasonally dry environments: the Mediterranean floras. Our gardens bear witness to the botanical wealth of these floras. The Mediterranean region itself, along with the nearby mountains of western Asia, were the original home to many of our favorite garden ornamentals. Daffodils, hyacinths, tulips, snapdragons, poppies, and oleander originated here. This same floristic area nourished western civilization with wheat, barley, lentils, peas, carrots, onions, lettuce, celery, parsley, artichoke, and many other vegetables and seasonings. Grapes, figs, olives, date-palms, and pomegranates also originated in this corner of the globe. Such a wealth of useful plants is a clear reflection of the generous diversity found within this flora. In fact, all five of the world's Mediterranean-type vegetation zones are included among Myers and Mittermeier's twenty-five biodiversity hotspots. Of these, one in particular has unusual plant diversity: it is the Cape region of southernmost Africa.[39]

SOUTH AFRICA'S CAPE FLORA

The flora at the southern end of Africa has puzzled botanists over many decades. Bordered by dry deserts and seasonally parched savanna woodlands, the Cape region supports a Mediterranean flora with winter rains and summer drought. The mountains are of old, eroded sandstone; soils are poor. Trees are few, and the landscape is mostly covered by scattered shrubs. Current estimates for the number of higher plants in the Cape floristic region stand at around 9,000 species, representing almost 1,000 genera and five endemic families.[40] For a seasonally dry flora covering only 74,000 km², these numbers are astounding. In a study to determine the global patterns and determinants of higher plant diversity, Holger Kreft and Walter Jetz subjected 1,032 regional floras to statistical analysis. Their worldwide analysis found that the Cape flora supports *twice as many species* as would be expected for a region with similar rainfall and temperature.[41] Why might this corner of the world, with limited rainfall and modest elevational range, be so rich in species?

The Cape flora of South Africa is particularly rich in irises, lilies, aloes, and orchids, but poor in annual plants. Many small- and stiff-leaved shrubs dominate this landscape. Succulents like the stone flowers, thick-leaved aloes, and cactus-like euphorbias are also numerous. In contrast, animal species are few in number, with only about 288 bird and 127 mammal species. Interestingly, over a thousand plant species in this flora have seeds dispersed by ants! Carrying the seeds into their nests—and feeding on the fat-body attached to the seed—results in having seeds deposited in a moister environment, where rodents can't find them. Also, since ants can't travel very far, this may have promoted local speciation. Along the same line, many of the Cape's plant genera have closely related species living in nearby but slightly different environments, suggesting that these species are the products of local ecological speciation.[42] Polyploidy appears to have been important in forming new species in a few lineages; the iris family with 295 species in the area has 27 polyploid species, nearly 10 percent.[43]

An interesting aspect of the Cape flora is that it has a number of genera with huge numbers of species. *Erica* of the blueberry family has 560 species in the Cape flora, while *Mesembryanthemum* numbered 700 species, before being split into segregate genera. *Aspalanthus*, a genus of leguminous shrubs, and *Pelargonium*, in the geranium family, each have more than 200 species in the region. Such species-rich genera are thought to have differentiated profusely over the last five million years. Using DNA analyses, Ben Warren and Julie Hawkins studied *Muraltia* of the polygala family, with over 100 species in the Cape region. They found that a majority of these species had formed over the last ten million years. That time period is in line with the origin of the Benguela Current, which brings cold waters from the southern ocean northward, and has produced drier conditions along Africa's southwestern coast. A diversification of *Muraltia* species over this time period may have been in response to that current's effects.[44] Perhaps the drying effect took place gradually, allowing plant populations to adapt and differentiate. And since we are invoking the factor of time again, perhaps we need to look more carefully at its role in biodiversity.

BIODIVERSITY AND REGIONAL HISTORY

In the last chapter, we used history to explain tropical Africa's poor floristic richness, postulating that severe and widespread droughts have eroded Africa's plant diversity. Regarding the Cape flora of South Africa, we can use this same reasoning to opposite effect. This is not contradictory. If severe drought—or other calamity—can cause species loss, then surely lack-of-drought will keep speciation going and result in higher numbers. Clearly, the southernmost tip of Africa has not been subjected to unusual species extinction over recent geological time. Given time, plant species have continued to subdivide the land amongst themselves. Unlike the flora of Australia, South Africa did not shift its position from one climatic belt into another. Southernmost Africa does exhibit a longitudinal species-diversity gradient, with the highest species numbers in the western sector. This is correlated with more topographic diversity in the west, and winter rainfall may have been more reliable in this area.[45]

Likewise, it is clear that Costa Rica has not suffered severe species loss in recent geological time. Growing on the trunks and branches of larger trees, epiphytes require a climate of reliable moisture for survival. Comparing two other regional floras, we found that Costa Rica had about twice as many dicotyledonous families of flowering plants with epiphytes than either Indonesia or the *Flora of Tropical West Africa*, far larger areas. Because epiphytes have disseminules that must be able to germinate on tree branches, such plants disperse widely. Many of Costa Rica's epiphytes may have migrated northward from the Andes. No matter the original conditions, a flora hugely rich in epiphytes implies a long history of reliable rainfall.

ENDEMICS AND LINEAGES: WHAT CAN THEY TELL US?

Historical biogeography has been richly informed by endemic species and lineages. These are plants and animals found only in a specific geographical area. Some may be recent evolutionary innovations, little more than curiosities. But others are hangers-on, survivors from more ancient times. The many endemics of Hawaii don't have much to say about ancient biotas; the islands are young, and many new species have originated there. The numerous fruit fly species in Hawaii are a recent and explosive radiation. A different story is related by the platypus in Australia, the tuatara in New Zealand, and the lemurs of Madagascar; these tell of a long and isolated history for the land masses where they survive.[46] Clearly, old islands can act as "shelters," harboring ancient lineages, while young islands can serve as "cradles," providing new opportunities for expansive speciation.

Mammals are especially informative in historical biogeography, and for two reasons. First of all, land mammals are not very good at getting across oceans. Compared to birds or the flowering plants, non-flying terrestrial mammals are tightly constrained in their travels. True, some mammals range widely over continents; the puma once roamed from Alaska to Patagonia, while leopards ranged from southern Africa to Southeast Asia. Nevertheless, land mammals cannot cross larger bodies of water. A second factor in their biogeographic importance is that mammals have a fine fossil record; we know a lot about the divergence times of their many families.[47]

Comparing the larger biogeographic regions, we find that 91 percent of Australia's mammal families are found nowhere else. South America ranks second, with 47 percent of its families endemic. Africa comes in at 36 percent, followed by the northern Holarctic Realm with 16 percent and the Oriental Realm with 13 percent. Looking more closely at mammalian geography, we must keep in mind that present geographic distributions can be misleading. Today, horses, zebras, and wild asses are only found in Africa and Eurasia. However, the fossil record makes clear that the ancestors of horses originated and diversified over many millions of years in North America. They spread into the Old World only later, where they are found today. Horses did survive into modern times in North America, only to go extinct after humans improved their hunting technology, around 12,000 years ago.

A fine example of an informative mammalian lineage is our own, the primates. Fossils of the earliest primates are from North America, about sixty million years ago. None of these ancient insectivores are alive today. A few other isolated and early branches of the primate tree do survive. The bush baby of Africa, the slow loris in Southeast Asia, and the aye-aye of Madagascar belong to these early primate branches. These are all small arboreal animals that are active only at night. The lemurs of Madagascar are another early branch; they represent a clearly defined larger element within the primates. Lemur-like animals are found as fossils in Europe from around fifty million years ago. Isolated from the rest of the world on their island home, lemurs retain many primitive primate features, including a more pointed, dog-like face. New World monkeys are considered more primitive than Old World monkeys because they've got more teeth in their mouths, and most lack the three-pigment color vision that characterizes Old World monkeys and ourselves. (More teeth are an ancient trait, better color vision is a modern trait.)

So there we have it. A variety of little primate oddballs in Africa and the Oriental realms. Lemurs confined to Madagascar, New World monkeys confined to tropical America, and the more modern Old World monkeys ranging from Africa and southern Europe all the way to eastern Asia and Wallace's Line. There's a pattern here. An important early phase of primate evolution survives only on the island of Madagascar, though similar animals once lived in Europe and Asia. New World monkeys survive only in the Americas, while the most modern primates are found over the largest contiguous area: Africa and southern Eurasia. Then there's Australia, with no primates at all, either today or in the fossil record. What do these broad geographic patterns suggest?

Lemurs represent an important stage in early primate evolution. They lived over a wide range in earlier times but are confined to Madagascar today. Similarly, New World monkeys are not the latest models in primate evolution, and they are no longer found in the Old World where their ancestors once thrived. The only reasonable conclusion is that—isolated on their island homes, whether Madagascar or South America—these animals were not confronted by the more advanced members of their own lineage, or predation by more modern carnivores. They have survived where they live by virtue of their isolation. And the even more ancient lineages represented today by lorises and bush babies? They survive in small numbers in deep forests, active only in the dark of night. The message of primate evolution is this: our world appears to have become increasingly competitive. Unless you are isolated from the action—or find a secure little niche—you may be vulnerable to extinction, as newer, higher-powered models take over the landscape. (We will return to the subject of "competitive escalation" in chapter 8.)

Having briefly reviewed the geography of species richness and patterns of diversity, we need to consider another fundamental ecological question. How is it that biological communities can sustain so many different species in the same place at the same time? Why haven't superior competitors taken over the landscapes for themselves? Why does so much of the world remain green, when there are literally millions of herbivores ready to devour all the greenery? How is it that complex ecosystems remain so complex?

Chapter 6

SUSTAINING LOCAL BIODIVERSITY

Wlive on a planet graced with many diverse ecosystems. Variety can be found at many sites, and at many spatial scales. We can enter a forest and look at the leaves on a branch and the insects feeding on that branch. We can also study the tree itself, or a stand of trees. Similarly, we can study biodiversity at the level of the local landscape, whether forest or prairie. Diversity at this scale has been called *alpha* diversity. In contrast, diversity over larger regions has been called *gamma* diversity. In this chapter, we will be considering how diversity is maintained at the local, or alpha, scale.

Many local biomes support large numbers of plants and animals living together, in what appear to be stable communities. Whether we are speaking of a tundra-like mountain top, a deciduous forest, or tropical evergreen forest, hundreds of species live together over what appears to have been thousands of years. This confronts us with some serious questions. Why haven't a few dominating species wiped out most all the others? Why hasn't the world's greenery been devoured by hungry herbivores? These questions have been central issues in ecology over many decades, and a variety of ideas have been put forward to understand the marvel of continuing local diversity.

THE CONCEPT OF THE NICHE

For zoologists, an answer to the puzzle of animal diversity has been the concept of the **niche**. Each species, it is assumed, has its own specific niche: an environment or lifestyle in which that species is most comfortable and most successful. Recall the two species of lice living on humans. One prefers to live amongst the hairs of our head; the other, apparently, requires the greater warmth and moisture of our private parts. By specializing on these very specific *niches*, two lice species share a larger environment (our hairy parts) without getting in each other's way. Similarly, four species of Hispine beetles feed on *Heliconia latispatha* plants in Costa Rica, but each species feeds on different parts of the plant.[1] (*Heliconia* is the banana-like genus with "lobster claw" inflorescences we see in floral arrangements.) These beetles are related species, living on the same host but with slightly different feeding preferences. We call this **niche division.**

On the savannas of Africa, we can witness a clear separation of hunting niches among the larger mammals. Here, lions are the largest carnivores, fully twice the weight of a leopard. Hunting in packs, they can bring down the largest herbivores, even elephants. The next smaller guild of carnivores includes leopards, cheetahs, hyenas, and hunting dogs. Each of these species hunt in a distinctly different way. Hyenas and leopards hunt mostly at night, hyenas hunt in groups and also scavenge, while leopards are mostly solitary and can climb trees to pursue monkeys. Packs of hunting dogs and solitary cheetahs hunt on the open grassy plains during the light of day. Using speed, cheetahs run down their prey; using relays, hunting dogs wear down their prey. Smaller carnivores also share this same landscape: the coyote-like jackal, the feline serval, and several species of mongoose. The desert cat of Africa is about the size of our domestic cat. Living in a dry environment with few prey, the desert cat is a solitary hunter. Like many other desert mammals, the desert cat conserves moisture by being active only in the cool of night. Clearly, Africa's carnivores achieve their diversity by coming in a variety of sizes and using different hunting strategies.

Ecologists tried to embellish the concept of the niche with precise and imaginative qualities. Every animal species, they proclaimed, has a "fundamental niche" in which that species prospers. Two species with the same fundamental niche must inevitably compete, they declared, with one eliminating the other. Others proclaimed that the niche existed as a "multidimensional hypervolume bounded by the species' tolerance factors for many different environmental factors."[2] Popular in the 1970s, these discussions produced few insights, nor did they help us understand plant life.

PLANTS AND THEIR NICHES

Botanists, in particular, had problems with the niche concept. Green plants all use the same basic resources—sunlight, water, and carbon dioxide—to build their tissues. Also, many species seem to grow together amicably, and it is often difficult to see where their niches differ. Ecologist P. J. Grubb suggested that, for plants, we should consider four component niches.[3] First is the **habitat niche**: the particular ecological factors necessary to each species. The second is the **life-form niche**: the size, three dimensional structure, and annual productivity of each species. The third is the **phenological niche**: determining the timing of seasonal growth, flowering, and fruiting of each plant species. Finally, the **regeneration niche** determines how the species replaces itself in the landscape.

Careful analyses of tree distributions within three neotropical rain forests recently discovered that about 40 percent of the species were associated with specific soil qualities. This study examined 1,400 tree species and ten essential soil nutrients in Panama, Colombia, and Ecuador. Even though most of the trees appeared to be similar in habitat preference, more than a third of the species had germinated in soil with specific nutrient characteristics.[4] Here is an aspect of their *habitat niche* we hadn't been aware of earlier. Similarly, though altitude proved to be the most important factor determining local tree diversity in Borneo, soil characteristics were also significant.[5]

Surely, one of the easiest ways to pack more species into a single habitat is by varying their sizes. Little cats, whether on the savanna or in the desert, are busy looking for little rodents; they are not competing with either lions or leopards. Also, cats are much too large to make insects an important item in their diets; bugs don't contain a lot of energy. Smaller mammals, predatory insects, and nimble little birds are the primary hunters of insects. And just as there are advantages to being large, there are advantages to being small. Smaller animals can accelerate quickly, manoeuver nimbly, and find many places to hide.[6] Best of all, they don't need a lot of food to keep themselves going. Similarly, little plants can live in places where the big ones can't. Epiphytes are a fine example. Leafy liverworts, small ferns, and orchids less than ten centimeters (four inches) tall can festoon the branches of trees in misty cloud forests. In deserts, short-lived ephemerals can sprout, flower, and fruit in a rainy season of only two months. For any species, size at maturity plays a big part of determining its role in the environment: the *life-form* niche.

Even when the same size, similar species can share the same landscape by timing their life activities differently: the *phenological* niche. A colorful example of differential timing can be seen in our Midwestern prairies. After a long and cold winter, growth on the prairie begins in early May. The first to flower are small plants, less than a foot (30 cm) high. These include lousewort, Indian paintbrush, toadflax, *Hypoxis*, and the white lady slipper. By June there's been more time to grow, and plants are two to three feet (60–90 cm) tall. These include golden Alexanders, the pink pasture rose, bluish lupines, and spiderworts. In early July, the prairie is in full flowering mode. This is when phlox, wild quinine, leadplant, prairie coreopsis, wild lilies, and many others are blooming. In August there are broad swaths of brilliant purple blazing star across the prairie, while flowering spurge sprinkles the view with little white inflorescences and early goldenrod adds a dash of yellow to the show. In August, and after sufficient rainfall, taller species reach over six feet (180 cm) high. By September, brilliantly yellow goldenrod species are going full blast, complemented by white and bluish asters and, hidden in the undergrowth, the elegant gentians. All the while, the less colorful—wind-pollinated—grasses and sedges have also gone through their species-specific sequence of flowering.

A prairie is rather like a grand stage on which different groups of dancers replace each other as the flowering season progresses. Perhaps this is more than the random seasonality of different species. If pollinators are in limited supply, sequential flowering reduces competition for pollinators, allowing more species to share the landscape. By avoiding competition through lifestyle adjustments, niche differentiation helps maintain local species richness throughout the world.

PLANT SUCCESSION

There's another important way in which plants divide the landscape along the gradient of time, but this is not within the cadence of a single year. Rip off the side of a mountain with a landslide, or scour a flood plain with a nasty flood, and you have bare land. Smaller weedy species can prosper in this open, sunny, windswept setting. These plants must be able to disperse their disseminules far and wide in order to "find" open space, and then grow quickly. Called **plant pioneers**, they are the first to colonize open ground. Once a short cover of greenery is in place, other seeds can settle in and produce taller plants. After a few years, the vegetation becomes tall enough and dense enough for the seeds of shade-loving trees to germinate and begin their lives. This process is called **plant succession**: the gradual transformation of bare ground to pioneering weeds and seedlings, then young forest and, finally, mature tall forest (where the rainfall is sufficient). In a sense, the pioneer plants *prepare the site* for the secondary invaders and these, in turn, set the stage for developing a "mature" vegetation.

After disruption and calamity, plant succession rebuilds the local vegetation. Insight into this process is evident at Peru's Manu National Park. Situated on the eastern flank of the Andes, and as these mountains continue to rise, the Manu River gradually gains in elevation, and its flood plain shifts. (The Andes Mountains are said to be rising "about as fast as your fingernails grow.") Consequently, older terraces are found along one side of the Manu River, as the river slowly moves eastward in response to increasing elevation. Because the shifting river is moving in only one direction, researchers can compare both younger and older floodplain forests adjacent to the river. Weeds and grasses are the first to colonize the river's edge. Soon, *Ceiba* and *Cecropia* trees, together with other early pioneers, begin their lives in the open sunlit greenery along the river's edge. As they begin to form a canopy, a new forest grows around them, reaching a stature of about 100 feet (30 m) within a hundred years. Plant succession has demonstrated what it can do with bare ground, and it doesn't stop there.

After two hundred years, this same tall forest exhibits the full grandeur we expect, rich in large, thick-trunked trees and vining woody lianas. By three hundred years, however, the early pioneers are collapsing; the forest is losing its stature. And by five hundred years, forest structure has changed dramatically. Most all of the early giants are gone, replaced by tall and spindly younger trees. Forest dynamics may have made a mess of the forest, but it has become an extravagant mess, richer in species with the passing of time.[7]

LIFE-HISTORY STRATEGIES

The temporal sequence of vegetation succession introduces us to another important niche dimension: life span. The little weed growing at the edge of a tropical river may be able to flower and produce seeds within two months. On that same river bank, a young *Ceiba* seedling may have just begun its life trajectory. If this plant can survive the tribulations of its riverside home, it may become a forest giant, producing flowers and fruits for over two hundred years. Some plants live quickly and die young; others bide their time, waiting decades before producing flowers and seeds. Dayflies may have only a few hours for their reproductive maturity, while elephants live for more than half a century. Biologists who study differing lifestyles call their subject **life-history strategies**.[8]

An especially revealing comparison focused on rabbits and small monkeys. Here we had two mammalian species of the same body mass, yet living very different lives. Rabbit females produce more offspring more often than do monkeys of similar weight. Fundamentally, rabbits aren't very smart, but they're *cheap to build*. Baby rabbits do not require as much energy and mothering as do baby monkeys. The reason is simple; rabbits have brains only half the size of monkeys of similar weight. Brains are expensive, requiring a lot of energy in construction, nurturing, and ongoing maintenance. With smaller brains, rabbit mothers have larger litters more often. The little monkeys give birth to only one or two babies, and these babies take much longer to mature. Especially significant in this comparison is *life span*. On average, these monkeys live more than twice as long as the rabbits! For the monkeys, a lower reproductive rate finds compensation in a longer life trajectory. Clever monkeys, living in treetops, produce fewer but smarter babies, and live a longer life. Rabbits have more babies more often, but with a shorter life span. Forged by natural selection, both strategies have proven successful over time.

How might natural selection have determined the length of life, both for us and all other species? The logic is quite simple. Once we—or any other species—reach an age where we no longer contribute to the well-being of our offspring, natural selection no longer applies. For example, if we carry genes that shorten our early lives and keep us from provisioning our children, those children are less likely to survive. Consequently, genes causing early-onset disease are less likely to be carried into the next generation. That's strong negative selection. However, if we carry genes making us susceptible to diseases when we are in our seventies and no longer helping our children or their children, such genes are *invisible to natural selection!* If the late-acting genes have no effect on the success of further generations there is no way they can be subject to positive selection! And thus we, and all the other species on this planet, are stuck with a life span appropriate to our lifestyle and our niche.

TROPHIC LEVELS

Niche differentiation can also be viewed from a broader perspective: that of trophic levels in the ecosystem. Here we have a pyramid of species, the base of which is made up by the **primary producers**—mostly green plants using sunlight to build new tissues. The next trophic level is that of herbivores, or **primary consumers:** species feeding directly on the greenery. The next level is made up of those species that feed directly on the herbivores; this crowd of **secondary consumers** includes carnivores, insectivores, and others. Finally, a third trophic level of parasites and pathogens feeds on all trophic levels. In addition, a large number of organisms consume dead organic matter; these include bacteria, fungi, and a variety of small animals. Taken together, these trophic levels form a pyramid, with each higher level having less biomass than those they feed on, since there is energy loss as one moves up the food chain. Nevertheless, higher tropic levels may have many more species than lower levels, as when many little herbivores are feasting on a single large tree.

Unfortunately, being at the top of the trophic pyramid can be precarious, as occurred when American populations of bald eagles collapsed. Aquatic plants had absorbed the insecticide DDT from water draining agricultural fields, and following mosquito-control operations. Little fish fed on these aquatic plants and accumulated DDT in their fatty tissues. Larger fish ate the little fish and further amplified DDT residues. Bald eagles, feeding on the larger fish, inevitably acquired high concentrations of DDT—and their reproduction fell drastically. Other raptors suffered similarly, making clear that DDT use had to be curtailed. Everyone had thought that pesticides would simply diffuse into the landscape and, to all intents and purposes, disappear. DDT contradicted this assumption, showing that physiological processes could sequester such chemicals, and the food chain could concentrate them, in a process called **biomagnification** or **bioaccumulation.** We humans are vulnerable to this same process with mercury, arsenic, lead, and cadmium, as we burn fossil fuels and mine for gold.

Niche division, together with a pyramid of trophic levels, gives rise to complex interconnected **food webs.** A recent study of aphids illustrates some of the complexities within a simple food web. Aphids are small (1–4 mm) insects that suck sap from herbaceous plant stems. You often find a cluster of aphids, busy sucking on a weedy stem, attended by ants. Since plant sap is deficient in some nutrients essential for aphid growth and survival, aphids have a problem. One solution is sucking up a lot of sap and sending most of it out their rear ends—while extracting the special nutrients they need. Because the sap they expel is rich in sugars, ants often tend and protect the aphids to get the sugary exudate. But aphids attract parasites as well as protectors. Little parasitic wasps attacking aphids are called parasitoids. The wasp injects an egg into the body of the aphid; this egg becomes a larva feeding on the insides of the doomed aphid, finally pupating and emerging as a new wasp. But things aren't really that simple. The parasitic wasps are themselves parasitized! These secondary parasitoids—also little wasps—come in two strategies: those that parasitize the primary parasitoid larva, and those that parasitize the parasitoid pupa. And that's not all—the host plants play a role as well. Using brussels sprouts and their wild relative *Brassica oleracea*, together with two aphid species and a number of parasitoids, this study showed how the quality of the host plants affected the numbers and diversity within this little network. Here the two species of aphids were subjected to five species of primary parasitoids, and these to ten species of secondary parasitoids. The result of these experiments was clear: the quality of the primary food affected the numbers and diversity in the aphid-parasitoid system. Turns out, the wild plants supported more aphids and a more diverse parasitoid community than did the brussels sprouts.[9] Since this study included only two plant hosts and two herbivorous species, imagine what's going on in rain forests!

DISRUPTION AND DIVERSITY

Niche differentiation may help us understand local species diversity, but it does not deal with the question we asked earlier: Why haven't superior competitors taken over the landscape? How is local diversity maintained over time? Why are there three hundred co-occurring tree species in some tropical rain forests? To answer these questions, biologists have looked at environmental variables that disturb the status-quo. These come in two general classes: physical phenomena such as storms, drought, and earthquakes; or biological factors such as predators, parasites, and pathogens. Let's begin with the physical.

While most tree species in a tropical rain forest are scattered about, it was discovered that some species were found in close groups. Not only were they the same species, they were the same age. For some clumps, the reason was easy to see. There was evidence of a landslide or large tree fall, with seeds or saplings of the clumped trees having been in the right place at just the right time to take advantage of the new clearing. Plants need sunlight for growth, and sunlight is in short supply on the floor of the rain forest. Any disturbance that allows more light to enter through a gap in the canopy will change the dynamics of the understory. In fact, some understory species are specialists on particular gap sizes. Three species of understory palms were studied in Costa Rica; all are specialists on a particular size of light gap. Too small a gap, and the palms do not have enough light to survive, but if the gap is too large other species overgrow them.[10] This is niche specialization for a very particular kind of forest disturbance. How might these palms manage to *find* the right tree gap, you ask? By producing a multitude of seeds and having a few of these many seeds germinate in just the right spot.

Whether the collapse of a giant forest tree, a river changing its course, or landslides opening new space, such processes disturb the status quo, allowing new immigrants to gain a foothold and begin their life trajectory. Driving through Central America, we were annoyed by the landslides delaying our progress along the Inter-American Highway. At first, I had thought: "Lousy engineers!" But after a few years of field work it became clear: the roads and the engineers weren't the problem, it was the mountains! These young, unstable mountains are constantly collapsing and producing landslides. In Central America, in the Andes, and along mountains round the world, ongoing geological uplift has created elevated highlands with steep unconsolidated soils. Throw in earthquakes, and you've got an unstable geology constantly opening up new ground, supporting high species turnover, and high species richness. Thank you, plate tectonics!

Physical disturbance is usually random; landslides, storms, or sudden tree-falls are a matter of chance. Most plants deal with this problem by constantly dispersing seeds; hopeful (so to speak) that a few will find themselves in an appropriate site. Some species play a waiting game. In carefully monitored Malaysian forests, it was observed that some immature understory trees, around thirty feet (10 m) tall, remained at that height over more than twenty years. Failing to grow further, these trees did not flower or fruit; they seemed to have a "wait-and-hope" strategy. If a nearby canopy tree does come down, our waiting tree will have the illumination needed to grow into the canopy where it can flower and fruit. Surely, unpredictable disruption is a major factor in supporting the species richness of many habitats, even allowing the lucky seed from a new species to join a plant community that is already populated by hundreds of other species! In a review of their studies of tree dynamics at Barro Colorado Island in Panama, Stephen Hubbell and Robin Foster concluded that "chance and biological uncertainty may play a major role in shaping the population biology and community ecology of tropical tree communities."[11]

Unpredictable variation in weather is another disruptive force helping keep plant communities diverse. One of the surprises of visiting our Midwestern prairies is how different they can appear from year to year. And it is not just rainfall; the severity and length of the previous winter can play a role. A very dry spring can alter the growth and flowering of different species. A good year for one species may be a bad year for another. For plants, the ability to produce seeds will rise and fall with varying environmental conditions. A recent prairie study in Kansas showed how inter-annual climate variability contributed to the coexistence of three very similar grass species. Seed storage in the soil, together with the overlap of generations—and climate variability—prevented any one of these grass species from dominating the landscape.[12]

Another important factor in our Midwestern prairies is *fire!* After witnessing a brilliant display of prairie flowering, I decided to share this phenomenon with friends. The following year, I filled my car with visitors and drove off to see the flowering extravaganza at the same prairie and the same time as the year before. Sad to say, but the flowering extravaganza wasn't there. There were flowers, of course, but nothing like the profusion of the previous year. My "Flower Tour" had nowhere near the impact that I'd been hoping for. Robert Betz, having studied these prairies over many years, noted that fires played a major role in the intensity of floral displays. When many nutrients are returned to the soil after a burn, the following summer is likely to produce a burst of flowering and fruiting. Similarly, fire is an important dynamic in many seasonally dry biomes around the world.

Moderate disruption is an important factor in maintaining local species richness. Of course, we need to distinguish moderate from severe disruption. Forests in Southeast Asia that are regularly subjected to destructive typhoons do not have as many tree species as forests outside the paths of these powerful storms. Similarly, a long cold winter or many months of drought are challenges only well-adapted species can survive.[13]

DIVERSITY PROMOTERS: PREDATORS, PARASITES, AND PATHOGENS

In addition to physical disruptions maintaining species richness, there are biological factors as well. After studying bruchid beetles in Central American lowland forests for many years, Daniel Janzen suggested that these beetles had a definite effect on the distribution of their host trees. Bruchids feed on the seeds of specific species of leguminous trees and are most numerous under such trees, "waiting" for fruit to ripen and seeds to fall. The beetles and their larvae feast upon the fallen seeds and destroy the seeds' ability to germinate. It was from such observations that Janzen and others proposed survival and regrowth of seeds to be *inversely proportional* to the distance from their mother tree. Seed survival and growth was more likely *further* from the mother tree.[14] Others have found that seedlings near a parent tree die off more readily than those more distant, because of pathogenic fungi in the soil surrounding the mother tree. Indeed, some tree species exhibit a so-called "hyperdispersed" pattern of occurrence within the forest.

But things aren't that simple. Seed-eating rodents, busy on the forest floor, seek out infested seeds with tasty grubs inside, adding another element to the story of seed success. Unlike the "clumped trees" we mentioned earlier, these trees of the same species simply don't grow near each other. Unfortunately, the seed predation hypothesis explains the distribution patterns of some species but not a whole lot of other species.[15] Nevertheless, specialized predators and pathogens can keep their host species from becoming too numerous. Even grazing animals can help maintain diversity. Careful studies in English meadowlands have shown that meadows grazed by sheep had twice as many plant species as those same meadows in the absence of grazing!

And there's more: competitive relationships are dynamic. We are all aware of the effect of predators on their prey, and we see the effects of **arms races** in many aspects of the living world. Horses are larger and run much faster today than their smaller ancestors of fifty million years ago. Many lineages of plants have developed specific toxins to defend themselves from herbivores. Though most herbivores cannot survive specific poisons, a few lineages have the ability to sequester the poison from the plant on which they feed, becoming poisonous themselves! Milkweed plants and monarch butterflies have such a relationship. Monarch caterpillars are one of only a few insect lineages that can feed on milkweeds (*Asclepias* spp.). Thanks to ingesting and sequestering the plant's toxin, both the monarch caterpillar and butterfly are quite poisonous, and both have distinctive, easily recognized colors. Over time, birds have developed the *inherited wisdom* to keep these butterflies and their caterpillars off their menu.

Plants of the rainforest and prairie may look like a pleasant arrangement of greenery, but they are much more than that. As in the milkweed example, plants are a veritable zoo of defensive chemistry. Cocaine in the coca leaf, caffeine in the coffee bean, morphine in the fruit wall of the opium poppy, and pungent aromatic oils in the mint family—all are defensive armaments against herbivory. Indigenous vegetation in

New Zealand has no plants poisonous to mammals because non-flying mammals were absent in New Zealand before the arrival of humans. Over many millions of years there simply was no need for mammal-repelling toxins in mammal-free New Zealand.

An important way of avoiding predation for animals is to hide. This is one reason for being active only in the dark of night. Beetles are clumsy fliers; they fly at night to avoid more agile birds. A majority of mammals restrict their activities to the cover of night. Another way of avoiding your predators is to mimic your surroundings. Grasshoppers look like blades of grass; many caterpillars stop when disturbed and appear to be a dead twig; moths fold their wings and mimic bark or fallen leaves; all are difficult to see. These creatures are the products of selection—deceiving the keen eyes of those who would devour them. Mimicry takes other forms as well. There are tropical moths that resemble stinging wasps; caterpillars with two large eyespots at their ends to look like little snakes, and brightly colored orange butterflies that look just like brightly orange *poisonous* butterflies. These many mimics bear graphic witness to the dangers of the world in which they live.

ENEMIES OF ENEMIES ARE FRIENDS

For plants, predators of herbivores are friends! A very dramatic example of predator effects on vegetation came with the reintroduction of wolves into Yellowstone National Park. By 1926, wolves had been exterminated in the park, with herbivore populations, especially elk, rising to new levels. Quickly browse species, such as cottonwood, aspen, willows, and berry-producing shrubs declined. And though coyotes, bears, and cougars were present in the park, they were not sufficient to constrain elk numbers (which had to be culled over the years). With reintroduction of thirty-one wolves during the winter of 1995/1996, the scenery began to change! Cottonwoods and willows in the Lamar Valley were now growing six to twelve feet high, instead of being continuously browsed down to three feet. Regrowth was especially strong near streams, where denser vegetation made it easier for wolves to hide and ambush their prey.

Smart elk were changing their browsing patterns, staying clear of denser growth where wolves might lurk. Soon the landscape began to resemble photographs taken in the early 1900s.[16] Yellowstone clearly demonstrated how top carnivores can help maintain higher levels of plant growth. All told, a dynamic habitat with a variety of predators supports greater plant species numbers. However, while large carnivores are highly visible factors in environmental dynamics, there are also many less visible "predators."

Sad to say, but the natural world is rich in smaller, often invisible, microbial parasites and pathogens. Such agents, from viruses to microbes and slender worms, all play a role in maintaining biodiversity. These deadly adversaries regulate the numbers of both animals and plants, and they are often constrained by simple physical parameters. If their hosts are numerous and crowded, the pathogens have little difficulty spreading to new hosts, causing epidemics. If the host species is rare or scattered, infection is less likely, and the parasite must bide its time. Here again, we have *density-dependent* factors keeping population numbers in check. This same process constrained human numbers before modern medicine. More important, pathogens sustain biodiversity by giving rare species a slightly better chance of reproducing successfully than their more common neighbors.

Young plants and animals are especially vulnerable to disease. In a study of tree seedlings on Barro Colorado Island, it was found that fungal diseases killed many of the little plantlets during their first year of life.[17] Earlier, we discusses how sexual reproduction kept a species' genetic resources in constant flux. In fact, incessant *genetic shuffling* by sexual reproduction may be the only way to deal with the challenge of parasites and pathogens. Sexually reproducing populations, by remaining constantly variable, give pathogens and parasites a less uniform target. And though we may abhor disease and pestilence (as well as storm and flood), all these factors are vital to maintaining rich and diverse ecosystems. Nevertheless, Mother Nature also has positive ways of sustaining biodiversity.

WHEN COOPERATION PROMOTES BIODIVERSITY

The notion that **cooperation** between different species might be a potent force in maintaining biodiversity has always been met with skepticism. After all, natural selection is supposed to work on the reproductive effectiveness of *individual* plants and animals. If that's true, how can helping others be a product of natural selection? Perhaps, the cooperation we see so prominently in nature originated independently in the cooperating partners. Ancient flowers, visited by pollen-eating beetles, may have reproduced more effectively than flowers not visited. Later, adding colorful petals, nectar and aroma attracted a greater variety of insects. Slowly, there developed a loosely symbiotic system, with insects finding food as plants were being cross-pollinated more effectively. These plants and animals had simply been maximizing their own benefits independently, before becoming enmeshed in a mutually beneficial syndrome. Regardless of who has benefited more in these interactions, the results have been hugely important in the evolution of more diverse ecosystems.

Animal pollination is the primary reason that tropical rain forests can support as many as three hundred species of trees in the same small area. Wind pollination can be effective, as we see in our oak woodlands, among savanna grasses, and in conifer forests around the world. However, it is effective when plant species are surrounded by many others of their own kind. Wind pollination is not very effective for isolated plants, such as in a tropical rain forest. Sentient animals can seek and find plants of the same species over long distances. Without sentient pollinators, a rain forest with hundreds of co-occurring flowering plant species would be impossible.

> *Insects serve only those plants that attract their attention, provide them with somewhere to land when they arrive, guide them in the direction they should go and reward them for their efforts.*
> —Peter Thompson[18]

But why might a pollinating insect search for the flowers of another plant of the same species? Actually, the underlying logic is simple: *conserve energy*! Why spend time trying to figure out how to suck nectar out of a new and different flower when you've just tanked up from a nice purple-and-yellow one? The strategy is clear: "That was good, let's find another purple-and-yellow flower." Even little insect brains can follow this protocol, resulting in what we call **pollinator fidelity**. In this way, pollinators tend to move between flowers of the same plant species. When the purple-and-yellow flowers run dry, the insect will initiate a new search.

The "loose symbiosis" of flowers and their animal pollinators has given rise to varying degrees of specific dependency. The bee family, with 20,000 species, is especially important. Bees are the only insect family feeding their young nectar and pollen. That's why you see honey bees and bumblebees visiting flowers so frequently. They are not just filling their own gas tanks, they are gathering sustenance for their little ones! And it's not just insects. Birds can also serve as pollinators, often attracted to bright red flowers. A signature feature of the New World tropics is the presence of bright red flowers with narrow floral tubes, and no landing sites! You will not find similar flowers among the native plants of the Old World. The reason is simple: hummingbirds are exclusive to the Americas. Because these little birds can hover in midair, they do not need a place to alight while sipping nectar. Pollinating birds in Africa, Asia, and Australia visit bright red flowers provided with nearby perches from which they sip nectar. No need for perches in flowers visited by hummingbirds. An estimated 8,000 species of flowering plants, in dozens of families, have evolved flowers specifically adapted to hummingbird pollination in the neotropics.

FIGS AND THEIR FIG-WASPS

An extreme—completely interdependent—pollination symbiosis is that of fig trees and their pollinating wasps. The fig tree's little flowers are enclosed within the spherical or urn-like fig. Though it looks like a fruit, the fig is actually a hollow inflorescence, with all its little flowers inside. Tightly overlapping bracts, covering the entry at the top of the fig, require that the flat-headed fig-wasp squeeze her way through this obstruction to enter the interior of the fig, usually losing her wings in the process. Once inside, she moves around among the female flowers of the fig with pollen collected from the fig in which she was born, thus effecting pollination. After laying eggs in many gall flowers, where her young will develop, her life will be complete. Once the wasp larvae have fed, matured, and pupated inside the gall flowers, blind and wingless males are the first to emerge. These raunchy fellows proceed to mate with the females, still trapped within their gall flowers. The males then get together and chew a passageway through the thick wall of the fig. Fresh air from this new opening activates the females, who emerge from their galls, gather pollen into special "pockets" alongside their abdomen, then leave by the exit the males have fashioned. The females must then fly to a fig tree that is both the correct species and whose figs are ready for pollination. This elaborate symbiosis includes three different types of flowers inside the fig. The female flowers come in two forms: the gall flowers in which the wasp larvae will develop and fertile female flowers, which will produce seeds. The third group are male flowers, producing pollen for transport by the female wasps. Soon after the female wasps leave, the fig becomes sweet and succulent, with seeds ready for dispersal by birds, bats, monkeys, and others.

Fig-wasps cannot develop or mate outside of the fig, and fig trees cannot produce seeds without the pollination services of the wasps. This is an **obligate mutualism**, a tightly constrained relationship essential to both members of the association. One might expect such a completely dependent mutualism to be risky business. If either partner goes extinct, so does the other. Indeed, the fig-tree/fig-wasp mutualism is one of only a very few such tightly integrated relationships, among tens of thousands of flowering plant genera. Nevertheless, the fig genus (*Ficus* in the mulberry family) is one of the most successful tree genera on the planet, numbering around seven hundred species. Fig trees have an advantage: they do not have to compete with other plants for pollinators. They've got their very own![19] And there's more: To keep themselves and their wasps going, fig trees fruit throughout the year in evergreen tropical forests.

This means that fruiting figs are usually available as food throughout the year, even during the dry season when other fruits are scarce. Consequently, figs are a major source of nourishment for monkeys during the dry season in Peru's Manu Park.[20] In this same forest, a single fig tree bearing an estimated 200,000 mature figs was monitored for twenty days. During this time, forty-eight species of birds and monkeys were recorded feeding at the tree during daylight hours. Considering how many bats and other creatures might come to feed at night, you can understand why fig trees often serve as **keystone species**, critical to the welfare of many other members of the forest community.

OTHER SYMBIOTIC ASSOCIATIONS

Ants, in particular, have developed many cooperative arrangements with plants. Small ants inhabit stems and leaf stalks in a few Central American species of *Piper* and small trees of the laurel family. These plants afford enclosed spaces that house the little ants. In return, the ants bring organic matter into the plant, supplementing the plant's nutrition.[21] Much more impressive are those plants that host a *police force* of larger, nasty, biting or stinging ants. Mess with these plants and you are met with a phalanx of little warriors ready to do battle. That's the good news; the bad news is that armies are expensive. Ant-plants must provide both housing and food for their defenders. The tree genus *Acacia* (in a wide sense) contains a few species, both in Africa and the American tropics that display this unusual mutualism. *Cecropia* trees, common in lowland wet forests of the Americas, support an army of stinging ants in distal hollow stems; only slow-moving sloths feed on these distinctive plants. After cutting down a slender *Cecropia* tree in Costa Rica to collect foliage and inflorescences, I stepped back and waited. I expected the ants to run over my boots and bite my legs, where I would pick them off. Unfortunately, these little beasts ran up my legs with such speed and determination that they didn't slow down to bite until they had reached a very tender region. I've never made that mistake again.

Another important symbiotic system resembles what we call agriculture. Both termites and some ant species fashion special chambers where they cultivate fungi upon which they feed. The leaf-cutting ants of the Americas bring leaf fragments down from high trees to provision their underground fungal colonies.[22] Tropical savanna termites keep their underground chambers cool by building hollow spires of dried mud high above the ground, allowing air to circulate within the colony. Because they concentrate nutrients and rework the soil, both termite mounds and leaf-cutter nests play an important role in many ecosystems.

Recent DNA surveys have given us a better understanding of a symbiosis critical to the lives of many animals. These are microbes within the alimentary canal that help digest food and form feces. Among larger animals, it is the ruminant ungulates (cows, buffalo, antelope, deer, goats, sheep, etc.) that have the most elaborate digestive systems. These animals have a four-part stomach in which chewed plant food is digested, regurgitated, re-chewed, and further processed along the alimentary canal before it is expelled. Bacteria, amoeba, and fungi are part of this digestive system, allowing these animals to make cellulose a part of their diet!

Even the little laboratory fruit fly (*Drosophila melanogaster*) has a number of bacterial species helping digest its food. These commensal bacteria are especially important to the fly's survival during times of food scarcity. We ourselves have a huge entourage of digestive assistants. Genomic screenings estimate that *several hundred kinds of bacteria, numbering in the trillions*, inhabit our alimentary canal, help digest our food, provide a few essential vitamins, and even defend us against diseases. This "symbiotic human microbiome" is critical to our own good health.[23]

Unfortunately, cooperation can also produce trouble. A nasty fungal disease of rice includes a bacterium living within the fungus. This bacterium produces a toxin (rhizoxin), killing the rice plant upon which the fungus then feeds. Thanks to its bacterial partner, this fungus can devastate one of our most important food crops. Another recent discovery involves a three-way partnership in a tropical panic grass. But this relationship is beneficial to the grass. Here a fungus, infested with a virus and living within the grass, confers greater heat-tolerance to the grass.[24]

On a larger scale, the most important symbiosis in terrestrial ecosystems is that of specific soil fungi joining with the roots of plants, called **Mycorrhizae** (fungus-roots; singular: mycorrhiza). In these symbioses, specific soil fungi attach their tissues to the outer cells of plant roots (Ectomycorrhizae), while other smaller fungi actually enter into the root tissue itself (Endomycorrhizae). Connected in this way, plant and fungus can exchange soil nutrients, water, and sugars. Fungi are especially good at absorbing available nitrogen and phosphorus from the soil, and they can transmit these elements to the root. Thanks to photosynthesis, the plant produces sugars in abundance, and these are available to the fungus through the mycorrhizal connection. Providing essential nutrients to the plant and energy-rich sugars to the fungus; this relationship is thought to benefit 80 percent to 90 percent of land plants. In fact, this symbiosis may have begun over four hundred million years ago, when plants were first establishing themselves on land.

Though not symbiotic, many fungi are critical in breaking down woody tissues and returning vital nutrients to the soil. Whether detritovores, pathogens, or symbionts, fungi are essential in sustaining healthy ecosystems. In addition, a few fungi have developed a very distinctive symbiosis, allowing them to live in even the Earth's most severe polar environments.

LICHENS: GLOBALLY SUCCESSFUL MUTUALISTS

Actually, we're not supposed to call them lichens any more. The appropriate nomenclature today is **lichenized fungi.** There are two reasons for insisting on a new label. The first is simple: most of the lichen is, actually, a fungus. The microscopic photosynthesizing algae within the lichen's tissue are minute and difficult to see. But the second reason is more fundamental: the lichens are not a single (monophyletic) lineage. Different fungi have formed lichen associations with different algae, and even with blue-green bacteria. DNA studies make clear that different fungi have formed lichen associations independently and at different times in the past.

Most remarkable, lichens are often the only visible non-transient forms of life in the world's most severe habitats. Lichens outnumber plants and animals by a wide margin in Antarctica. Mosses are found in a few sites, with higher plants represented by less than ten species in Antarctica. Only a few species of birds and mammals survive there, feeding on the riches of the southern sea. At the other end of the Earth, lichens are a prominent part of the northern tundra, helping nourish herds of caribou. Lichens also flourish on high mountains and rocky out-crops exposed to wind and weather. We often see them on tree trunks and rock surfaces. North America boasts about 3,600 species, ranging in color from brilliant orange and bright yellow to quiet gray and dull brown.[25] Some tropical lichens look like dabs of paint on smooth tree trunks. Though not that conspicuous amidst all the greenery, six hundred species of lichens have been found within one square kilometer (0.38 sq. mi.) of lowland rain forest in Costa Rica's La Selva preserve.[26]

What makes lichens so adaptable? Surely their most important quality is that they are **autotrophic**; they can feed themselves thanks to their photosynthesizing partners (called photobionts). Safely tucked away within their fungal host, the photobionts are nurtured and pro-tected. Though the fungal element remains constant in a species of lichen, the algal element may vary. That is, the same lichen species may have different photobionts in different parts of its range. The algae provide carbohydrates to the fungus through "leaky walls." In turn, the fungal body of the lichen has the job of remaining firmly attached to an appropriate substrate, while resisting the wear and tear of wind and weather. In addition, the fungus provides the alga with water and min-erals. Simply put: "Lichens are fungi that discovered agriculture" (as lichenologist Trevor Goward once said). Numbering more than 17,000 species, lichens are a splendid example of biological cooperation.

LEGUMES: ANOTHER MAJOR SYMBIOTIC SYSTEM

Another grand symbiosis, helping support biodiversity in many ecosystems, is that of legumes and rhizobial bacteria. The legume family (Fabaceae) includes many impressive tropical trees, from flat-topped acacias of African savannas to giant rosewood trees in rain forests. The family also includes many twining plants, like garden peas, wisteria, and kudzu. Legumes provide us with nutritious foods: chickpeas, lentils, kidney beans, peas, soybeans, peanuts, and many others. Complementing our grain crops (such as wheat, barley, rice, and maize), the legumes, or pulses, are an important source of amino acids, essential for building proteins. Not only are the legumes more nutritious, they add *available nitrogen* to the soil when they die and decompose, contributing to sustainable agriculture in this way. By supporting nitrogen-fixing bacteria, legumes enrich themselves and the soil in which they grow. But why is this symbiosis so special?

Molecular nitrogen is unavailable to plants and animals, despite being a major component of our atmosphere. The reason is simple: molecular nitrogen has two nitrogen atoms in tight embrace, locked together by three strong covalent bonds. Only a few bacteria have the chemical machinery that can rip those two nitrogen atoms apart, transforming the relatively inert nitrogen of air into available nitrogen for other living things. Since amino acids, proteins, DNA, and RNA all contain nitrogen, available nitrogen is essential for all living things. But there's a problem: these nitrogen-fixing bacteria abhor oxygen! And this is why legumes produce pink-colored nodules on their roots, in which these bacteria live and work. The nodules are pinkish because they contain a special hemoglobin, which can bind oxygen, maintaining an oxygen-poor zone where the rhizobial bacteria can rip apart molecular nitrogen!

The legume symbiosis is a great story, but there's a problem: benefits have costs. Building nodules and nurturing nitrogen-fixing bacteria require a significant expenditure of energy. This means that a field of legumes is less productive of seeds than is a field of grains. Likewise, wild legumes living in natural habitats need to expend energy on maintaining their symbionts. Nevertheless, the sacrifice has been grandly successful. Numbering around 19,500 species in over six hundred genera, the legume family contributes significantly to the world's biodiversity.

A few other plants and animals harbor nitrogen-fixing bacteria. A remarkable example is found in termites, where the bacteria that help termites digest cellulose can also fix atmospheric nitrogen in a three-tiered system.[27] These bacteria live only inside protozans that live in the termite's gut! With bacteria on board that can both digest cellulose and fix nitrogen—and with water from their own respiration—termites are able to prosper in dry dead wood.

The big message here is that, just as endosymbiosis created eukaryotic cells, and intercellular cooperation made multicellular organisms possible, positive interactions have also fostered greater complexity within modern ecosystems.[28] A striking aspect of these many important biological mutualisms is that they are members of different kingdoms: algae within fungi forming lichens, legumes with bacteria to fix nitrogen, flowers and the animals that pollinate them, microbes helping us digest our food.[29] Such examples make clear how "getting along well with others" can be a sound survival strategy in a difficult world. But having reviewed cooperation, disruption, disease, and niche division in maintaining local biodiversity, we need to examine another fundamental driver of biodiversity.

AVOIDING EXTINCTION: REPRODUCTION AND DISPERSAL

Cruising down the Amazon during an ecotour, it was my job to give lectures in the evening. I had just finished talking about biodiversity and it was time for questions, when a tour member seized the moment. An author of business management texts, this gentleman was sharp, as was his question.

"What purpose do mosquitoes have?" he demanded to know.

"Making more mosquitoes!" was my response.

Every organism on this planet has the same intrinsic drive: not just survival, but **reproduction.** Whether it is a solitary mushroom sending millions of spores into the wind, or a primate mother carefully nurturing her single offspring, the bottom line is the same for everybody. Just as death ends the life of a multicellular individual, extinction of a species means the game of life has ended for that particular lineage. In the real world, natural enemies, natural disasters, and plain old bad luck are part of the landscape. Fail to produce viable offspring and—in a biological sense—your life has been for naught. Here again, **natural selection** is the driving force. Healthy individuals with a strong reproductive drive will produce many offspring. Unhealthy individuals, or those who aren't interested in sex, will produce few. With these traits governed by multiple gene interactions, a strong reproductive drive will be maintained in every species, automatically, and for all time!

Every proper plant and animal has the same set of inherited instructions: *reproduce!* Those that fail this dictate compromise the future of their species. Reproductive protocols are central to the behavior of every living thing. Nutrition and survival may be the primary parameters, but they only serve the urge to procreate. This urge, in turn, powers both local and global biodiversity. For many lineages, the best strategy for long-term species survival is producing lots and lots of offspring. Year after year, elm trees cover our lawns with tens of thousands of little winged seeds. Cottonwood trees waft away countless cottony seeds on the winds of early summer. Spawning fish release millions of eggs and sperm into lakes and streams. Producing huge numbers of fertilized eggs or seeds is one way to keep the threat of extinction at bay.

In botany we often call seeds and fruits **disseminules**; their job is moving on out. Mobility is a central feature of survival. For animals, mobility allows escaping or avoiding predators, or finding better forage in a distant landscape. For plants and fungi, it is the ability to send seeds and spores far and wide across the landscape. With a bit of luck, a few of these disseminules will find the right niche in which to grow and flourish. In times of climatic change or local calamity, mobility may determine which species survives and which cannot. In the marine world, many animals have minute larval stages that allow the tiny offspring to disperse far and wide. More significantly, these huge numbers, whether animal larvae or fruits and seeds, are a primary food source for others in the community, helping sustain local biodiversity.

Today, both deliberately and accidentally, humans are dispersing many species around the globe. Consider Ascension Island, far from other landfalls in the mid-Atlantic ocean. When Darwin visited this island, the vegetation was treeless, and he promoted the introduction of woody plants. Now, over 150 years later, the island is richly covered in vegetation, with "cloud forests" contributing to stream flow. Here, deliberate dispersal has assembled a new self-sustaining ecosystem.[30]

Large numbers is one way a species can play nature's lottery of life, but there are other, equally effective, strategies as well. Having fewer offspring but generously tending and carefully protecting them can also foster reproductive success. Elaborate behaviors such as burying, hiding, or encapsulating the little eggs increases survivorship. Birds prepare a safe nesting site, then diligently rear a new brood each and every spring. Likewise, mammals invest virtually all of their adult lives in protecting and nurturing their young, generation after generation. Many land animals, and even a few fish, have abandoned egg-laying altogether. They may produce eggs, but the eggs remain within the mother's body to develop before the young are released into a hostile world. Some cichlid fishes are "mouth brooders," holding the little ones within their mouths, providing the little fry protection until they become larger and better able to fend for themselves. Overall, two very different reproductive strategies are obvious: (1) make lots of eggs/spores/seeds and disperse them into the environment, versus (2) give birth to fewer offspring and take good care of them. For this latter group "species do not just happen, but must be *achieved* in each new generation, held in the world through the labor, skill and determination of individual organisms in real relationships of procreation, nourishment and care."[31]

Many species exhibit additional "insurance policies" with which they improve their survival over time. Bacteria can transform into spore-like states to survive difficult conditions. Many species of plants and insects cannot survive dry weather or long winters; rather, their little offspring must endure these difficult episodes in the form of seeds, eggs, and pupae. A further strategy in the business of survival is to have seeds or eggs of the same season vary in their **dormancy** periods. Even under the best conditions, some seeds or eggs will remain dormant the following year, "waiting" several seasons before they sprout or begin developing. If next summer brings severe drought after all your seeds have sprouted, they will likely perish, and you have failed to reproduce your kind. Delayed dormancy, on the part of a percentage of your offspring, can avoid such a disaster. Many animals do the same. In fact, some soybean pests now remain dormant in the soil, as farmers grow two years of corn over a three-year crop rotation in the same field. You guessed it: these soybean pests have "learned" to cycle with the farmer's crop-rotation pattern. Actually, the pests hadn't learned anything. But, with variable and genetically determined dormancy times, it didn't take too many generations to have selection get most of the pests cycling in synchrony with the rotation of their soybean hosts. By staying dormant over two years, these pests emerge for the third year in the rotation: soybean time! Thanks to environmental selection, Mother Nature has solved a lot of problems in the business of survival. Best of all, staggered germination times give the plant community a **seed bank** sitting in the ground over many years, ready to emerge after stressful times. However, while surviving minor calamities is one challenge, being fed upon by nasty neighbors is yet another.

DYNAMIC INTERACTIONS!

Predators, parasites, pathogens, and herbivores have given rise to a cruel and dangerous world—though one with millions of species. There have been "arms races" between every species and its enemies over time, making the local environment a very harrowing and dynamic arena. Predators are being selected to become more effective; pathogens are

evolving to counter the defenses of their hosts. Large numbers of herbivores play an important role in maintaining vegetation with many co-
occurring species.[32] A host of ongoing ecological interactions, both negative and positive, explains how local communities can support so many
plants and animals. Having a specific niche in all this turmoil is one way
to survive. But this is also a playing field in which *good luck* plays a significant role. And, because good luck is largely a random statistical operation, many lucky species can stay in business within diverse and complex
ecosystems. All the while, rampant reproduction and effective dispersal
keep the system going.

In his book *The Diversity of Life*, Edward O. Wilson referred to the ESA
theory of biodiversity.[33] ESA stands for Energy, Stability, and Area. **Energy**
for life, of course, comes mostly from the Sun, and this factor explains
why life is more abundant in the Sun-drenched tropics. **Stability** contrasts with major cataclysms, such as expanding ice sheets, extensive
drought, or sudden climate change. Finally, a larger **Area** will support
many species and this can be critical after a local region suffers annihila-
tion. Migrants and propagules from the larger region can recolonize the
locally devastated area and rebuild the local biota. As species go extinct
in one place, they can be replaced by migrants from elsewhere. A steady
flow of new immigrants requires similar communities over a larger area,
and this is why area is so important in Wilson's formulation.

Today, natural communities are becoming smaller in the face of
human expansion; sources of new native immigrants are diminished,
and replacement becomes less likely. At the very same time, and thanks
to global commerce, species from distant regions, arriving without their
pests and pathogens, may become explosively invasive. Whether a fungal
disease exterminating the American chestnut, zebra mussels devouring
phytoplankton in the Great Lakes, or West Nile virus threatening birds
and ourselves, we live in a world of increasing extinction (a topic we'll
revisit in the final chapters).

Coevolution is a crucial process in the organization of biodiversity specifically because it is simultaneously flexible yet conservative in how it shares interspecific interactions in a constantly changing world.

—John N. Thompson[34]

Nature is dynamic, both globally and locally. Unreliable weather, nasty parasites, harrowing diseases, fearsome predators, and a few helpful mutualists, all interacting in complex environments, help sustain diversity in the same area over time. From a local perspective, all evolution is coevolution. Darwin used the example of an "entangled bank" to remind us of species interacting at a local scale. Often encountered along streamsides, light from above the open water illuminates a tangle of shrubs and clambering vines along the bank. Maintaining local biodiversity is an operation of many entangled and interacting factors. All this is contemporary ecology, helping us understand today's biodiversity. But how did such elaborate biomes arise? Pursuing that theme, let's move on to paleontology, examining the escalation of biodiversity on planet Earth across deep time.

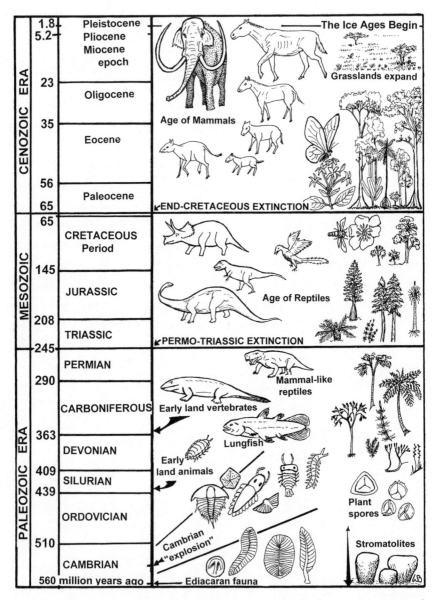

Figure 1. The last 560 million years of Earth's history. The Phanerozoic eon. Epochs are listed for the last 65 million years. Periods are listed for the time between 65 and 560 million years ago. Note that the time-scale for 0 to 65 million years ago is not the same as the time-scale for 65 to 560 million years ago.

Chapter 7

THE EXPANSION OF BIODIVERSITY ON PLANET EARTH

Meteorites are rocks that have, quite literally, fallen from the sky. Using the isotopic ratios of a variety of elements within these meteorites, astronomers estimate that our solar system formed around 4,560 million years ago (mya), and that our planet solidified just a little later.[1] Those must have been perilous times, as a solar system full of cosmic debris slowly organized itself into the stability we enjoy today. Craters covering both the Moon and Mercury are a reflection of heavy asteroidal bombardment in those early times. In fact, some planetologists claim that the giant gaseous planets, Jupiter and Saturn, with their significant gravitation, have vacuumed up a lot of solar system debris over these last four billion years. Larger comets and asteroids are still out there, some with erratic orbits that threaten us, but they are few and far between.

Astronomers believe that bombardment by comets and asteroids was especially intense around four billion years ago, bringing with them much of the water that makes our planet so distinctive. Just as significant, a large impact may explain our Moon's beginnings. According to the **Big Whack hypothesis,** a Mars-sized object collided with our planet to become one of the *really lucky breaks* in Earth's history. Thanks to this calamity, Earth ended up with greater mass and a more powerful magnetic field, a field acting both to deflect energetic particles fired at us by the Sun, and reducing the erosion of gases from our outer stratosphere. Best of all, by restraining our planet's axial wobble, our large satellite has given us more stable terrestrial climates.

IN EARLY TIMES

The early history of our planet has been divided into three main stages, called eons. The earliest is the **Hadean Eon:** 4,560 to 4,000 mya. This was indeed a hellish time, as our solar system was still convulsed by meteoritic mayhem. The second stage is called the **Archean Eon:** ending around 2,500 mya, when iron deposits began precipitating in oceans around the world. The third was the **Proterozoic Eon** (before animal life): 2,500 to 542 mya. The oldest rocks at the bottom of the Grand Canyon reach down into the Proterozoic; they bear no fossils of larger living things. Over the first 4,000 million years of Earth history, our planet gives no evidence of larger plants or animals. But there are larger objects, probably constructed by bacterial activity, in those ancient times.

The first larger fossil structures formed by living things are called **stromatolites.** Quite common in the early fossil record, they are rare today; found only in a few shallow tropical marine embayments. Stromatolites are often rounded cushion-like or columnar rocky structures, generally one to two feet (30–60 cm) in diameter and a few inches to six feet (10–200 cm) high. They can also form as flat layers in shallow water. What makes them distinctive is their thin, laminated structure. In living examples, microscopic layers of bacteria and algae live near the surface, trapping sand and debris in their gelatinous membranes. With time, new layers are formed and trap more debris, creating a layered cross-section. The earliest stromatolites are dated to about 3,500 mya.

Many geological processes can give rise to laminated structures. Cone-like stalagmites often have a layered structure due to chemical changes in the water from which they are formed. Wave action along shorelines can also produce thin-layered strata of differentiated sediments. However, the lamination within fossil stromatolites suggests they were built by thin layers of bacteria, together with trapped sediments. Rare today, they are found only in warm, highly saline water, where aquatic grazing animals are absent. Stromatolites, however, were common along shallow tropical sea shores around the world—until the advent of predatory animals.

Fossil evidence of larger organisms is first found around 560 mya, in Ediacaran times (635–542 mya).[2] These organisms are mostly leaf-like devices, attached to the ground at their stalked-base. Others are tubular, and some are radial with smooth edges, resembling pancakes. Others had parallel transverse ridges, like little sleeping bags. Most are the size of your hand, but a few leaf-like forms grew to a meter (3.3 ft.) in height. None were more than an inch (2.5 cm) thick. None had external apertures or appendages. Clearly they couldn't eat anything and appear to have grown by adding segments at one end. More importantly, they were immobile, having left no trails. Also, there is no evidence of anything having taken a bite out of them. Numbering around fifty well-characterized species at this ancient time, I suspect that they were autotrophs, nourished by Cyanobacteria enclosed within transparent jellyfish-like tissues. Found only as impressions in ancient rock, we have no clue to their internal or cellular structure.[3]

Many Ediacaran sites were inhabited by numerous individuals of only a few species, just as is the case in some cold-water habitats today. A cold-water habitat may also explain why these fossils seem to show little bacterial degradation. Indeed, the world before Ediacaran times had suffered massive glaciation. Geologists believe that land surfaces had congealed into a supercontinent around a billion years ago, called **Rondinia**. Blocking ocean currents, this grand landmass caused our globe to suffer two major ice ages. Some even suggest that the most severe of these episodes resulted in a planet covered with ice: the **Snowball Earth** episode. For living things, this would have been a harrowing time, surviving only in the ocean's warmer depths. Worst of all, an ice-covered Earth, reflecting warm sunlight, might have remained frozen. However, with volcanic activity injecting CO_2 into the atmosphere, the planet warmed again, thanks to plate tectonics!

The enigmatic Ediacarans were followed by a period of small, fragmented shelly fossils. This period, around 545 mya, is when fossil trails and burrows first became common. Such was the fossil evidence for larger life forms during these early times. Then, quite suddenly, life propelled itself forward.

ANIMALS ARRIVE: THE "CAMBRIAN EXPLOSION"

There is nothing like the Cambrian until the Cambrian.
—Andrew Knoll[4]

Between about 542 and 530 mya, the ocean world exploded with larger forms of life. As just noted, rocks around 545 mya exhibit an increase in burrowing activities and trace fossils, together with a mix of small, shelly fossils. Then, about 535 mya came the **Cambrian Explosion** itself. The earliest stage includes the mollusks, brachiopods (lamp shells), and echinoderms (starfish and sea urchins). By 530 mya, the fossils include trilobites and the ancestors of animals, which would later give rise to scorpions, horseshoe crabs, and fish. Within a period of less than ten million years, the rocks suddenly include the remains of many animals bearing heads, tails, legs, spines, eyes, and other useful features. Most of these new creatures left an indelible mark in the record of the rocks for a very simple reason: they had a hard, resistant covering. Surely a crowd of hungry neighbors was the reason for growing a tough fossil-friendly epidermis.

Most surprising, the Cambrian explosion of animal fossils features a wide range of animals with fundamentally different body plans. There are no "proto-animals" that gradually diversify into more varied forms. Rather, the Cambrian explosion is marked by an array of very different creatures from its very beginning. Both coiled and clam-like mollusks, a variety of worm-like animals, segmented trilobites, primitive crustaceans, chelicerates (ancestors of spiders, mites, and horseshoe crabs), starfish-like echinoderms, and weird forms with no modern counterparts are all here and all together. As the Cambrian curtain rose, the stage was already populated by a variety of differing lineages.

Paleontologists have been puzzling over the mystery of this sudden and dramatic event for over two hundred years. The Cambrian period had been preceded by extensive ice ages, perhaps even a *Snowball Earth*, and the enigmatic Ediacarans.[5] However, this was well before the explosive appearance of so many different forms of animals appearing in grand synchrony at the beginning of the Cambrian. The discovery of minute animal fossils, less than a tenth of an inch (2.5 mm) in size, in rocks around 560 million years old, indicates that many ancestral lineages were already there—but they were miniscule! Why might these different lineages have expanded in size during so small an interval of time?

The powder—as represented by the little fossils—was there; but what lit the fuse? What caused the sudden increase of size in separate animal lineages at this one moment in time? It makes sense that if the Earth had gone through a severe cold period, only the smallest forms could survive, protected in small areas of warmth within the ocean. But why suddenly diversify and expand between 542 and 530 mya? Only one scenario makes sense: a major change in the world's environment allowed different animal lineages to become larger at the same period in time. Whether these creatures built their hard parts of carbonates or silicates or phosphates, all appeared during this same short time interval.

The simplest and most plausible theory for the Cambrian explosion claims that life itself had changed the world. Beginning much earlier, blue-green bacteria had been busy splitting water in photosynthesis. Perhaps a billion years later, eukaryotic algae added their efforts to this life-affirming activity. Using the hydrogen atoms of water to fix carbon and build energy-rich compounds, photosynthetic Cyanobacteria and algae *released free oxygen into the atmosphere.* Ongoing oxidation reactions with various other elements would purge the atmosphere and oceans of free oxygen over many millions of years. However, sometime around 2.5 billion years ago, oxygen levels began to rise.[6] A striking aspect of our planet's crust is that nearly all the world's banded-iron formations were deposited between 2,500 and 1,800 mya. Thanks to an atmosphere rich in carbon dioxide, huge amounts of ferrous iron were held in solution in the oceans. As free oxygen became more abundant, ferrous iron reacted with oxygen to precipitate as ferric iron deposits in a grand "rusting" of the planet! Today these ancient banded-iron formations represent about 90 percent of our industrial iron reserves.

AN EVOLVING ATMOSPHERE

Earlier, the Earth's geology was mostly gray, but now, with oxygen doing its work, reddish rocks give evidence for "The Great Oxidation Event" around 2.2 billion years ago. Life itself had begun to change the atmosphere. Nevertheless, oxygen remained at very low levels, resulting in what paleontologists call the "boring billion" years. It wasn't until between about 635 and 548 mya that isotopic ratios in sulfur compounds give evidence that oxygen concentrations were rising to modern levels.[7] **Oxygen pressure**, in all probability, was the key factor in having so many different kinds of larger animals all show up in the record of marine fossils *at the same time*. Powered by oxygen-consuming respiration, larger animals require more oxygen. It was not until oxygen became a larger portion of the atmosphere that larger active animals were able to evolve. Here was the most sudden escalation of complexity in the history of life. An earlier world, devoid of larger animals, was now alive with nearly all the major animal phyla we know today.

The Cambrian explosion may have been the opening salvo in the proliferation of larger animals, but it could not have been their true beginning. Studies of DNA relationships among the major animal phyla, extrapolating back in time, suggest that these lineages had differentiated much earlier.[8] How else might Cambrian rocks contain such a wealth of *different lineages* of animals? They had originated while still tiny and were already evolving along independent trajectories. As oxygen levels rose, and the firepower for energizing larger animals became available, different lineages all responded in a similar way: producing a diverse array of much larger forms. Suddenly, the environment became more diverse and much more dangerous. The "boring billions" were long gone, a world with larger, fiercely competitive, animals was a completely new world.

The Cambrian period began suddenly, with animal lineages continuing to expand and diversify over time. The Cambrian was followed by the Ordovician (488 to 443 mya), a period marked by a grand increase in species richness but terminated with a sudden loss of many species. The

Silurian (443 to 416 mya) and the Devonian (416 to 360 mya) periods followed. Each of these periods was characterized by suites of dominant marine animals, and each transition was marked by faunal collapse or change. Overall, the number of animal genera increased from around seven hundred in the late Cambrian (500 mya) to fifteen hundred in the Ordovician (450 mya), then fell, only to return to higher levels over the next two hundred million years.[9] During these early times, the sea was alive with fascinating creatures that have left a rich fossil heritage. The first bony fish are found around 500 mya, but they lacked jaws; lampreys and hagfish are their living descendants. Fish with jaws began their own grand radiation about 416 mya. Around that same time, another profound advance in life's history was underway.

PLANTS PIONEER THE LAND

> *Land plants changed the world of processes as surely as they changed the world of places.*
>
> —Oliver Morton[10]

The grand procession of life we've just reviewed was entirely marine; moving onto the land was a more formidable challenge. We find the first evidence of microscopic **spore-tetrads** in the fossil record around 470 mya. Nothing living in the sea produces such spores; they represent something new and different. These spore-tetrads appear to be the reproductive disseminules of the most simple, and most ancient, land plants. For the ensuing fifty million years, these spores are the only evidence we have regarding the possible presence of land plants. No stems, no leaves, no roots—just spores. Sad to say, small plants decay quickly, leaving little for the fossil record. Spore-tetrads are four spores stuck together in a single unit. Recall meiosis, where two cell divisions produce four haploid sex cells. The spore-tetrad is the direct product of meiosis, with the four haploid daughter cells remaining stuck together. These spores have tough moisture-retaining walls, allowing them to survive long dry periods and travel widely. Built with a tough covering of sporopollenin, spores readily became fossilized. (Fungi and bacteria also produce spores, but these are quite different in form and chemistry.) Spores-in-tetrads are the first evidence that plants were adapting to a terrestrial environment.

Just as higher oxygen levels were essential to the metabolism of larger animals in early Cambrian seas, increasing oxygen concentrations was important in the advance of life onto land. Higher concentrations of oxygen gave rise to a thin layer of ozone (O^3) high above the atmosphere. This high-elevation **ozone layer** acts as a shield surrounding the globe, absorbing much of the ultraviolet (UV) radiation the Sun beams our way. Because high-energy UV light can damage living tissues, the ozone layer plays an important role in our biosphere. Think of what strong sunshine can do to unprotected light-skinned people. For creatures adapted to living in water—where UV light is absorbed—the ozone shield helped make the land environment less hostile.

Surely, blue-green bacteria, various algae, and a variety of fungi were already living on moist land surfaces many millions of years before the arrival of green plants. Today, they form the dark green borders you see at the muddy edge of ponds, and the yellowish green glop you find in moist depressions. These are algae and bacteria still doing what they've been doing for a very long time. But such communities flourish only with continuing moisture, and they quickly become inactive when desiccated. Overall, they don't amount to much more than slop and glop. In contrast, the first terrestrial plants probably resembled thin prostrate little **liverworts**. Still with us today, liverworts are restricted to moist forests and shaded muddy soils. (Together with mosses, liverworts are members of the Bryophyta.) Liverworts come in two general categories: flat, strap-shaped green forms, half an inch (1–2 cm) wide, growing close to moist substrates. Today, *Marchantia* is found around the world. In contrast, the leafy liverworts have very small leaves (a quarter inch / 2–5 mm, or less), borne on slender stems that rarely exceed three inches (8 cm) in length. Leafy liverworts are often found hanging from branches in the moist understory of wet forests. Lacking erect stems, and with thin tissues, these kinds of plants make poor candidates for fossilization. In all probability, liverwort-like plants produced the earliest fossil plant spores, leaving no fossils of themselves.[11]

Around 425 mya, the fossil record changes: spore-tetrads decline in abundance, and solitary spores become common. Bits of plant tissues begin to show up in the fossil record as well. Land plants are beginning to make changes. **Lignin** to strengthen tissues made of cellulose was an important innovation. In addition to reducing water loss through their outer surfaces, a waxy layer protected against ultraviolet radiation, microbial attack, and corrosive chemicals. Unfortunately, having a waxy covering presents a problem: how can a plant absorb carbon dioxide from the air if it's covered with wax? Moist interior cell surfaces must be exposed to air in order to absorb carbon dioxide for photosynthesis. And that raises yet another problem; exposing moist cells to air will result in water loss by evaporation. Because carbon dioxide concentrations are low in air, a great deal of water will be lost in the process of absorbing carbon dioxide. **Stomates** were the solution to this problem—little openings on the surfaces of the plant. Called the stomatal apparatus, or stomate for short, the little openings have two peripheral guard cells that can open and close in response to moisture stress. Opening for gas exchange and closing to conserve water loss made stomates an essential land plant innovation.

Unfortunately, gas exchange results in water loss, and larger plants must replace water with their roots, and this means a larger plant needs plumbing! Tubular cells within the stem die and lose their contents. Connected at their open ends, and with their insides empty, these dead cells become conduits, allowing water to move upward unobstructed. We call such tissues the plant's **vascular system**. With a vascular system, water absorbed from the soil can be carried up to the photosynthesizing tissues, replenishing the water lost through evaporation and used in photosynthesis. Combined with cellulose, lignin helped create stronger vascular tissues and stiff stems. In strong contrast, simpler liverworts lack lignin, stomates, roots, and plumbing, which is why they are both small in size and restricted to wet environments; their tissues must absorb moisture directly from their surfaces.

By around 420 mya, we find the first fossil plants with erect stems reaching 4 to 8 inches (10–20 cm) high. Their simple green stems carried spore-producing structures at their tips. Though lacking leaves, they did have a vascular system, able to transport water up from below and keep their green tissues moist. Because their roots were small, these plants remained short and slender. Nevertheless, and for the first time, the terrestrial world was becoming a lot more interesting.

PLANTS GROW LARGER

With the development of stomates, lignin, plumbing, and roots, land plants moved into high gear. Plants grow with meristems: embryo-like tissues that can continue growing to form new leaves and stems. Meristems are generally found at the tips of stems and roots, where they continue building new tissue. (Plants grow by the continuing growth of their meristems, unlike animals, which lose their embryonic tissues during early development.) Around 400 mya, plants invented something new in the way of meristems. **Cambium** is a thin tubular meristem surrounding the center of the stem and paralleling the entire long axis of the stem. A single cell layer thick, the cambium produces new cells on both its inner and outer sides. In this way, the cambium forms new vascular tissue, causing the stem to expand in girth. Becoming just a little thicker and a little stronger each growing season, the cambium builds new plumbing as well as additional structural support. By creating strong wood fibers toward the interior, cambium produced woody tissue along the entire length of the stem. With cellulose reinforced by lignin and a long tubular cambium, plants began building trees.

> *The whole point of a tree is to colonize the air, where sunlight is plentiful.*
>
> —Diane Ackerman.[12]

The earliest trees were leafless. Rather, they looked somewhat like palms, with a solitary trunk and shorter dissected greenish branches at the top;

branches that would fall to the ground as the tree grew taller.[13] Not until around 370 mya did trees begin building leaves. Thin leaves, with broad flat surfaces to intercept sunlight, are a more efficient way of harvesting light, but they require more plumbing and elaborate branching for support. As time progressed, different lineages developed a variety of tree forms. By around 365 mya, moist tropical estuaries and river basins were the home of forests! Soon, trees were reaching as high as 80 feet (24 m), while fern-like plants graced the understory. All this greenery was a major advance for the world's biodiversity. In fact, around 300 mya, giant millipedes and dragonflies with two-foot wingspans, enlivened these ancient forests (probably because of higher oxygen levels at that particular time). Overall, the Carboniferous period (363–290 mya) was marked by widespread forests in river basins and estuaries. These were the times during which massive coal deposits were formed (the primary energy source for our electrified society).

As in so many other situations, positive advances on one front produce negative effects in another area. The spread of forests during the Carboniferous had a global effect. Vast quantities of CO_2 were removed from the atmosphere and ended up in dead vegetation, eventually becoming coal. Because CO_2 acts as a thermal blanket for our planet, this period of vegetative exuberance was followed by the most extensive glaciation of the last 580 million years. Deposited at roughly the same time period around the world, coal deposits mark a significant episode in our planet's long history. But why wasn't all this organic matter degraded more rapidly? Nowadays, termites, some beetles, and many fungi are decomposers of wood, preventing our forests from becoming piled high with dead wood. A recent study, combining fungal phylogenies and fossil dating, claims that lignin-digesting "white rot fungi" first appeared around three hundred million years ago.[14] Using enzymes that can tear apart lignin, these wood-decaying fungi may have put an end to the Carboniferous!

Land plants are another dramatic example of how the living world has changed over geological time. Life on land—both plant and animal—initiated the grandest advance in the history of biodiversity and ecological complexity on our planet. But what might have induced any plant or animal lineage to abandon a moist aquatic home to live in a far more stressful environment on land?

THE CHALLENGE OF LIVING ON LAND

Why might plants and animals have adapted to a hostile land environment, when remaining in a comfortable liquid medium seems much the simpler choice? With natural selection operating only in the here-and-now, adaptations for some possible future environment are unlikely. The most reasonable answer to this question is that adaptation to life on land actually began by trying to stay alive within an aqueous environment. Both plants and fish living in shallow inland waters were often confronted by periods of desiccation, as their creeks, marshes, and mud puddles dried up. Many tropical regions have long and severe dry seasons during which no rain falls. Rain forests are the exception; they make up less than 30 percent of the tropics' potential natural vegetation. Most tropical regions are seasonally dry, ranging from deciduous woodland and grass savannas to hostile desert.

Tropical regions with limited rainy seasons have many rivers, swamps, and marshes providing rich aquatic habitats. A majority of these same wetlands lose their standing water over the dry season. Such conditions create life-threatening challenges for local aquatic plants and animals. Today, East African lung fishes burrow deep into the mud as their watery home dries up. There, within a mucous lining and living on stored fat, they can breathe air and survive until the rainy season returns. Long ago, lobed-fin fishes used their forward fins to move through shallow streams blocked by tree-falls and aquatic vegetation. With time, these vertebrates have come to dominate the terrestrial world.

Similarly, land plants arose from fresh-water environments. Comparative DNA studies make clear that the liverworts and their allies are most closely related to a group of living algae found only in fresh water. For plants, adaptation to the land probably began by growing quickly during wet periods in shallow pools and streams, and producing spores that could survive desiccation or be dispersed to other moist sites. The notion that the regularity of ocean tides and a rich seashore environment was the springboard from which animals became adapted to the land may hold true for many invertebrates, but this was not the case for either land plants or terrestrial vertebrates.

A variety of different animals, in fact, crawled out of their watery world and became adapted to life on land. The aquatic ancestors of spiders, insects, millipedes, a variety of different snail lineages, worms, and the land vertebrates (tetrapods) all became terrestrial *independently*. Each of these lineages came from a different aquatic phylum, not closely related to the others, having made the same journey onto land. In strong contrast, all land plants appear to have originated from a single lineage of complex freshwater green algae: the Charales.[15] Freshwater habitats, bordered by moist soils, provided the environment from which plants could pioneer the land. Lacking the movement and versatility of animals, plants had a far more difficult task in adapting to the land environment and, it appears, only one such attempt was successful. Also, and unlike most of the land animals, land plants acquired an entirely new body plan as they made the transition. Beginning with a larger diploid plant body—something quite rare among the algae—this lineage also developed an embryo and new growth modalities. With these innovations in place, plants began their invasion of moist tropical lowlands.

To summarize, early land plants probably lived along streams and ponds that dried up during part of the year. Their first adaptation was the microscopic spore or spore-tetrad, which could survive severe drying and be scattered across wide areas, germinating—if lucky—in a distant pool of quiet water or moist mud. Spore-tetrads are the earliest evidence of land plants. Waxy cuticles, stomates, and a vascular system were later innovations. And, when the tubular cambium came on line, woody stems fashioned a more diverse and complex vegetation.

Land plants proved to be globally transforming, not only as a major advancement of complex life, but also for increasing the numbers of living species. Plant cover intercepted the impact of rainfall, reducing erosion. The dead remains of plants added organic matter to the soil and more nutrient run-off into the sea. A leafy plant cover provided shade, protection from desiccating winds, and a rich source of food for herbivores. By absorbing sunshine and evaporating moisture into the air, plants reduced temperatures under a hot tropical Sun. Within a three-dimensional land flora, animals had new opportunities for relentless diversification.

SEEDS: ANOTHER INNOVATION FOR LAND PLANTS

Living descendants of the early vascular plants include the club mosses (Lycophyta), as well as the horsetails, ferns, and their allies (Pterophyta). Today, the lycophytes and horsetails rarely grow more than a few feet tall, but during earlier times some members of these lineages grew to become tall trees. Living ferns range from a few inches in height to over fifty feet tall. Slender tree ferns with broad and graceful leafy fronds are found only in wet tropical forests today, resembling plants that lived 300 mya. Together with giant lycopods, horsetails, and smaller ferns, tree ferns helped create ancient swamp forests. Among the giants in these ancient forests, *Sigillaria* grew to over 90 ft. (30 m) and *Lepidodendron* reached 130 ft. (40 m). Entombed in rivers and deltas, these forests formed the great coal deposits of the Carboniferous period. As these early forests were flourishing, a new plant innovation came upon the scene.

The development of the **seed** marked a major advance. In a way, the seed represents for plants what internal fertilization and a tough desiccation-resistant amniote egg represented for land vertebrates: reproduction without water! Even today, liverworts, mosses, ferns, and most amphibians require external moisture for fertilization. Frogs and toads call loudly from ponds and streams in early spring; they can only make love in a watery environment. Their sperm cells must swim through water to reach eggs deposited by the females they have embraced. Thanks to internal fertilization, reptiles can engage in sex without having to find a river or pool in which to couple, and a leathery egg meant that their young did not have to develop underwater. Reptiles soon spread far and wide across the landscape. Similarly, the invention of pollen gave seed plants an alternative to swimming sperm cells. **Pollen,** resembling tough little spores, can travel many miles without becoming desiccated. Carried by the wind—or by animals—pollen can reach the **ovule,** where the egg cell resides, and effect fertilization. In contrast, ferns, mosses, and liverworts require thin films of water for their sperm to reach an egg cell. Just as in the case of animals, these sperm have long wiggly tails that propel them toward the chemical signals of the egg cell.

Carried by the wind through the air or transported by animal agents, pollen grains emancipated seed plants from the necessity for water as the medium for fertilization. Germinating near the ovule in response to chemical signals, pollen grains split open to produce a microscopic tube that grows toward the egg cell. Carrying the male nucleus to where it can unite with a female nucleus achieves fertilization—all without having liquid water at hand. The fertilized ovule then develops into a seed, which can be dispersed to form another plant. Seed plants were now able to propagate in a variety of terrestrial ecosystems where water was scarce. Seed plants (Spermatophyta) included the extinct seed ferns, the living cycads, conifers, ginkgo trees, the strange gnetophytes (*Ephedra,* *Gnetum,* and *Welwitschia*), and the flowering plants (Angiosperms or Magnoliopsida).

The triumph of the seed was two-fold. First, pollen grains did away with the necessity for liquid water in fertilization. Secondly, both pollen and ovule simplified the two-stage life cycle of more primitive plants. Technically speaking, the original haploid male plantlet has transformed itself into the pollen grain! Similarly, the female haploid plantlet became a small part of the ovule produced by the diploid mother plant. With the proper chemical messaging, a pollen grain germinates at the ovule, grows toward the egg cell, and effects fertilization, and the resulting diploid embryo matures within the seed, ready to create an adult (diploid) plant.

FLOWERING PLANTS:
ANOTHER ADVANCE FOR BIODIVERSITY

The emergence of seed plants was followed by yet another major innovation in land plant evolution. In rocks dated around 140 million years old, we find the first good evidence for flowering plants or Angiosperms: their distinctive pollen and a few fossils. (Because plants decay rapidly, many plant lineages have left only a meager fossil record.) Expanding rapidly during the latter half of the age of dinosaurs, the flowering plants have been instrumental in creating more complex ecosystems. Today, flowering plants dominate most of the world's land surfaces. Of a conservatively numbered 300,000 species of land plant, flowering plants have over 260,000 described species.[16] Best of all, Angiosperms come in a huge variety of shapes and sizes. Giant baobab trees, oaks, alders and acacias, potatoes and barley, water lilies and roses, cacti, grass, and orchids are all flowering plants. Enormous diversity in size and structure mark the Angiosperms. But what is it that makes the flowering plants so different?

The critically defining feature of flowering plants is that their seeds develop within a tight enclosure: the **ovary**. Somehow, the flowering plant ovule (or a group of ovules) became enclosed within one or a few leaf-like structures. These leaf-like devices (called carpels) formed the ovary, within which the ovules are protected, pollinated, and nourished. As the ovary matures into a fruit, each fertilized ovule becomes a seed. This new device—*the ovary ripening into a fruit*—is one of the glories of our modern world. The mature ovary gives us the juicy tomato, the tasty papaya, the nutritious avocado, the woody coconut, and the pod that houses peas. Enclosed within additional floral tissues we have the apple, the cherry, the pumpkin, and the watermelon. In a few, the floral axis gets involved, forming the strawberry and tropical cherimoya, while the inflorescence axis helps produce a pineapple. In other lineages the ovary wall remains thin and hard, a simple encasement for the seed. This group includes our most important source of nourishment: the cereal grains of the grass family (Poaceae). Low in moisture, high in nutrition, and easy to store, cereal grains allowed humans to build grand civilizations. Wheat, rice, maize, barley, sorghum, rye, oats, and the millets continue to power the human enterprise today. The seeds of legumes or pulses (Fabaceae), with higher protein contents, also played a central role in sustaining human communities. These include peas, lentils, chickpeas, beans of many kinds, soy, and peanuts. Some societies have used tubers, corms, and roots as primary food sources. These include potatoes, yams, sweet potatoes, cassava, carrots, turnips, and others. All these food plants are Angiosperms! And just as the flowering plants have powered the human enterprise, they are the primary energy sources for a majority of terrestrial ecosystems. By diversifying so grandly themselves, flowering plants allowed other terrestrial lineages to diversify as well.

What was it that gave Angiosperms the ability to prosper in such a grand variety of shapes and sizes? The answer to this question involves another unique aspect of flowering plants—called **double fertilization**. After the pollen grain splits open, germinating on the stigma of the flower, it forms a pollen tube. This microscopic tube grows toward the ovules, carrying two or three nuclei. Upon reaching the egg apparatus inside the ovule, one of these nuclei will unite with the egg nucleus to effect fertilization. So far, nothing special: not much different from sperm nucleus joining egg cell in so many other plants and animals. But there's more: *a second nucleus from the pollen tube also gets into the act!* By joining with two free nuclei within the egg apparatus, the ovule begins to build endosperm—the tissue that will nourish the developing embryo within the seed. Called "**double fertilization and triple fusion**" this advance in plant complexity fostered a major expansion of biodiversity.

THE ADVANTAGE OF DOUBLE FERTILIZATION

Does double fertilization, yielding a triploid endosperm with three sets of chromosomes, make much difference? For the developing embryo and seed, probably not. Nevertheless, double fertilization was a key factor in Angiosperm success. To understand the importance of double fertilization, we need to look at that other grand crowd of seed plants, the Gymnosperms. These include pines, cedars, redwoods, and other conifers: cycads, gingkoes, and the strange Gnetopsids. **Gymnosperm** means *naked seed*, because the seed is not enclosed within an ovary. **Angiosperm** means *hidden seed*, since it forms within the ovary. Before the arrival of flowering plants, Gymnosperms dominated terrestrial landscapes, and they still do in the grand conifer forests of America's Pacific Northwest and Rocky Mountains, *Podocarpus* forest just beneath the summit of Mt. Kenya, or the vast needle-leaf forests of the northern taiga. But despite a long history, they number less than a thousand living species, and all are woody. This is an important point: there is no such thing as an herbaceous Gymnosperm! In striking contrast, Angiosperms have herbs scattered all across their many lineages, from lilies and grasses to petunias and an awful lot of weeds. What explains the ability of flowering plants to make so many weeds?

Gymnosperms just aren't very clever. Most of them produce seeds and endosperm *in advance of fertilization*. Not smart! Wind-pollinated Gymnosperms depend on frequent winds and closely grouped populations to effect fertilization; when these parameters are lacking, fertilization may fail. Producing energy-rich seeds that cannot sprout—because they weren't fertilized—is a huge waste of energy.[17] Larger woody plants can afford this waste; little plants cannot! To repeat: most Gymnosperms produce energy-rich seeds before fertilization has taken place. That's not the way Angiosperms operate. What **double fertilization** does is to initiate endosperm production only after fertilization has occurred. Thanks to double fertilization, precious resources will not be wasted on seeds that cannot sprout. In addition, tight coupling of fertilization with endosperm initiation speeds up seed production. This is why flowering plants have been able to produce so many little short-lived herbs and why, after three hundred million years, the Gymnosperms have never built a weed.

FLOWERING PLANTS: GREATER COMPLEXITY
YIELDS GREATER DIVERSITY

Flowering plants have achieved greater morphological complexity than any other plant lineage, and they've done this in many ways. In chapter 3 we mentioned how chromosome doubling can result in forming new species: this same process has played a major role in flowering plant diversification as well. Angiosperms have a remarkably resilient genome; adding to, or doubling, their genomes results in few negative consequences. Within the flowering plants, individual genome sizes vary over a more than two-thousand-fold range! At the bottom end sits *Genlisea* in the bladderwort family with a 63.4 Mbp (mega base pairs) genome, while *Paris japonica* in the lily alliance has a genome around 130,000 Mbp. Compare this with mammals having a five-fold range between their smallest and largest genomes, or birds with a meager two-fold range.

Best of all, flowering plants elaborated yet another clever tactic: *advertising!* Many flowers embellish their sexual parts with a bright whorl of colorful petals, enticing aromas, and sugary nectar. A great variety of animals respond to this enticement to fill their gas tanks, carrying pollen from flower to flower. **Colorful flowers** probably originated in different lineages and at different times over the history of the flowering plants. The *Magnolia* flower has a long central axis bearing spirals of stamens and ovaries (rather like leaves on a stem). However, the magnolia flower is very different from either the cup-like flowers of the laurels, the three-parted flowers of the lilies, or the five-parted flowers in wild roses and many others.

Animal pollination was a major advance in our planet's ability to support high species numbers. Wind pollination is effective for species that are well represented in the local landscape. With lots of other individuals of your species nearby, some of your pollen is likely to reach a stigma for pollination. However, if your numbers are few and scattered across the landscape, wind pollination becomes unlikely. There is no way in which wind pollination could sustain a modern rain forest with three hundred different tree species within a few acres. In a modern rain forest, many species are represented by only one or two individuals per acre—not enough to make wind-pollination work effectively. Whether imbedded in a species-rich rain forest, or scattered across desert thorn bush, brightly colored flowers are visible to their pollinators. Only sentient agents can travel from one bright orange blossom through dense underbrush or across a rocky landscape to find another orange bloom. Pretty flowers (most of them) offer something important to their animal visitors: sweet energy or nutritious pollen. By making rarity less likely to result in extinction, **animal-mediated** pollination gave the Angiosperms a big advantage. Starting with perhaps only a few hundred species 120 mya, and with over 260,000 species today, flowering plants have grandly expanded ecosystem complexity over the last hundred million years.

Let's return to the structural variety of Angiosperms. Their shapes range from cacti that look like spiny barrels to many-branched shrubs, tall trees, elegant palms, and weedy herbs. This profusion has transformed the world. Unlike a conifer forest with its many narrow spires, and thanks to wood with greater tensile strength, Angiosperm forests have canopies with widely spreading branches. This means that there is more to grow on (if you are an epiphyte) or swing from (if you are a monkey). Also, flowering plants have more specialized vascular tissue, both in the water-carrying xylem and the food-dispersing phloem. Leaves and leaf-architecture vary widely among flowering plants. Slender leaf-stalks (petioles) not only allow the blade to orient itself for greater exposure to the Sun but also provide flexibility in gale-force winds. In addition, leaf-stalks can form abscission layers at their base, to jettison the leaf at the end of a growing season. While not unique to flowering plants, **deciduous leaves** allow trees and shrubs to survive in strongly seasonal environments. Also, many Angiosperm leaves have high vein-densities, allowing them to transpire more moisture, while capturing more carbon dioxide and photosynthesizing faster.[18] By pumping more moisture from the soil into the high canopy—where most of it is evaporated—flowering plants *sustain the rain forest*. This is why cutting down a rain forest changes local climate so drastically.

Angiosperm variety has resulted in a spectacular range of differing fruits and seeds. Sizes range from coconut palms and cannonball trees (whose falling fruit might crack your skull) to the powder-like seeds of orchids. Many fruits provide succulent tissues that are eagerly devoured by animals: from the many-seeded watermelon to the single-seeded avocado. Others lack fleshy tissues but harbor nutrients, as in the cereal grains. Many fruits lack food value but are built to travel with the wind, outfitted with winged extensions or cottony fluff. Others have hooked hairs to latch onto passing animals and clothing (the inspiration for Velcro). Together, fruits and seeds help disperse flowering plants far and wide.

So much variety in growth forms, fruits and seeds makes clear why other plants and animals have increased their numbers in concert with the expansion of the flowering plants. Modern ferns expanded their numbers shortly after the flowering plants began diversifying.[19] Ants also began proliferating around 100 mya.[20] Beetle lineages feeding on flowering plants have many more species than do related beetles that do not use flowering plants as their food source. Epiphytic plants and small animals have also increased their numbers as Angiosperm-dominated forests expanded.[21]

Though these radiations began around 100 mya, the flowering plants have continued innovating in more modern times. Grasslands and savanna woodlands are among the most extensive modern biomes. They appear to have begun expanding about 30 mya, as indicated by the increasingly deep (hypsodont) teeth of large grazing herbivores. Since grasses do not fossilize easily, we infer the expansion of grasslands by the increased length of teeth among grazers. Silica crystals in grass cells wear down the teeth of grazers; these herbivores responded by growing longer teeth over evolutionary time. Grasses can regrow from their base after both fire and grazing. (This is why you can mow your lawn but not your petunias.) Decreasing temperatures over the last thirty million years have lowered rainfall and increased fires over many regions, allowing grasslands to expand.

More recently, a number of grass lineages have pioneered a new kind of photosynthesis, called C_4 photosynthesis. Not only do these grasses seem to have a competitive advantage in hot dry environments but some burn hotter. In this way, they erode woodlands more aggressively, widening their domain with each successive fire. From our own point of view, expanding grasslands with their large nutritious grazers provided the setting in which we tripled our brain volume over the last three million years! All told, flowering plants have been the primary drivers in building today's terrestrial biodiversity.[22]

RULERS OF THE LAND: THE TETRAPOD VERTEBRATES

As we've already noted, land vertebrates evolved in freshwater rivers and estuaries. In fact, much of their preparation for becoming creatures of the land developed while they were still really fish. Living in small streams and estuaries, these fish faced the challenge of moving through water obstructed by fallen logs and aquatic vegetation. Negotiating such obstacles required the help of their front fins. Turns out the ancestors of land vertebrates had fins arising from projections on the side of their bodies. With a lobe-like base, the front pair of fins had greater freedom of motion to push their way forward. Soon, the base of the fin became elongated—the origin of arms. Later, these arms were articulated with elbows, and they developed digits for traction. Living in waters laden with silt and poor in oxygen, these animals had already developed lungs to gulp air—augmenting the work of their gills. Abandoning the buoyancy of water, a rib cage supported larger lungs for a more dynamic life on land, as their hindquarters developed a second pair of legs. But what might have driven such a unidirectional trajectory? Most likely, hunger! The shores of shallow streams were home to tasty prey; none of the early land vertebrates were vegetarians. Around 360 mya, four lateral fins were being transformed into four legs, which is why most land vertebrates are **tetrapods**.

Vertebrates had a number of important advantages over their invertebrate neighbors. By carrying one's architectural support inside, there wasn't the problem of having to shed a thick outside armature when growing larger. Also, an internal skeleton allowed for a greater variety of muscle-attachments and enhanced flexibility. Bony fishes had elaborated the many-parted backbone, providing both rigidity and some flexibility. By propping up this backbone with two pairs of limbs, land vertebrates developed what William K. Gregory called "the bridge that walks." The tetrapod backbone resembles a suspension bridge; anchored fore and aft by shoulders and hips. From a waddling amphibian of 340 mya, with arms and legs splayed out sideways (as were the fins), legs became positioned under the body for more efficient movement in more advanced lineages. With the vertebral column as central axis, evolution fashioned a variety of designs, from giant herbivorous dinosaurs to graceful antelopes and the speedy cheetah. By losing legs and multiplying vertebral elements, Mother Nature also fashioned slithering snakes. Becoming two-legged (bipedal) was another innovation, allowing carnivorous dinosaurs, their descendants the birds, and human hunters to become so successful.

A strong internal architecture allowed land vertebrates to diversify into a variety of forms, from frogs to pheasants. With spacious lungs, a diaphragm to pump air, and a strong four-chambered heart, land vertebrates could grow large and remain active. Fishy scales were transformed into a variety of dermal coverings, most elegantly in the feathers of birds, and comforting in the fur of mammals. For over 150 million years, both carnivorous and vegetarian dinosaurs ruled terrestrial environments. During that time our own lineage, the mammals, remained small and mostly nocturnal. Being active at night made hearing an important aspect of survival. Three bones in the jaws of reptiles became transformed in early mammals and are now part of the inner ear, allowing mammals to hear a wide range of frequencies. Better hearing, in turn, gave rise to larger and more versatile brains, as warm-blooded mammals processed more information, more quickly.

Without question, the history of organic evolution has been marked by the emergence of new and more complex body plans. The elaboration of multicellular organisms, more than 550 mya was a major leap forward. Later, the land biota gave rise to ever-more species and lifestyles within terrestrial habitats of ever-greater complexity. But it wasn't all honey and roses; serious extinction events have also marked the history of life.

EXTINCTIONS: CHAPTER ENDINGS IN THE HISTORY OF LIFE

Early geologists trying to piece together the history of our planet found evidence for sudden changes in fossil faunas over short intervals of time, and they used these changes to demarcate different periods in the geological record. Continuing discoveries did not bridge these major discontinuities. The Cambrian explosion of complex animals was real; these animals were not preceded by smaller versions of simpler creatures. A sudden decline of diversity in the late Ordovician was not an artifact of insufficient exploration or fewer outcrops. For earth scientists, drastic biotic changes were useful, helping define chapters in life's long history. Extinction events were used to separate time periods. Of all these abrupt changes, paleontology's biggest extinction ended the Permian period, about 250 mya. It is estimated that over 80 percent of marine species became extinct during this prolonged calamity—the **Permo-Triassic Extinction**. Huge outpourings of lava in Siberia at exactly this time, poisoning air around the globe, seem the likely cause.

A second major punctuation mark in the history of life, the **End-Cretaceous Extinction**, took place quite suddenly 65 mya. All around the world, and at what appeared to be the same moment in time, the fossil record changed. Dinosaurs, the grandest animals that ever roamed the surface of our planet, suddenly went missing.[23] Only one dinosaur lineage, the feathered birds, enlivens our modern world. Giant reptiles of the sea—the mosasaurs, plesiosaurs, and ichthyosaurs—also vanished. Foraminifera, minute marine animals, collapsed in great numbers, and only slowly recovered. Not as severe as the Permo-Triassic extinction, the end-Cretaceous event was more recent in time and seemed more abrupt. With less evidence lost to erosion over time, this extinction could be studied in greater detail.

Why so sudden a demise for so many extraordinary animals? This question remained a back corner of evolutionary studies until a small group of scientists proposed a very dramatic scenario. In the June 6, 1980, issue of *Science* magazine, Luis Alvarez and his associates claimed that an impact of extraterrestrial origin had wiped out the dinosaurs! You don't get much bolder than that in the halls of academic science. As evidence, the Alvarez team announced their discovery of an excess of iridium in the fine-grained clay layers marking the end of the Cretaceous period. Their evidence came from sediments in northern Italy, Denmark, and New Zealand, where this specific time period was clearly preserved. What they discovered was a sudden increase in iridium concentration at the very end of the Cretaceous. Knowing that iridium is much more common in meteorites than it is in the Earth's outer crust, the Alvarez group claimed the "iridium anomaly" was evidence for a devastating impact by a large asteroid!

The asteroid-impact hypothesis initiated a frenzy of activity, both by those hoping to support the idea and those trying to demolish it. Further work in Scandinavia, the western United States, and Australia made clear that the anomaly was indeed worldwide and had apparently occurred at precisely the same moment in time. The discovery of shocked tektites (bits of glass and fused particles produced by a high-energy impact) were also found around the world at this particular moment in time. With the largest tektites found in and around the Gulf of Mexico, and using deep-core samples from oil-drilling operations, scientists have identified the buried crater itself, the Chicxulub structure, beneath the edge of Mexico's Yucatan peninsula.[24] Over time, a clear consensus was reached: the iridium anomaly is the signature of a catastrophic encounter with an asteroid.

Extensive volcanism, with huge amounts of lava erupting in India at around the same time, also stressed biotas at this time. But the so-called "bolide impact" seems to have been crucial. Exactly how a meteorite between 2 and 6 miles (0.8–10 km) in diameter might affect the Earth on impact continues to be the subject of speculation. Smashing into the Yucatan's limestone rocks would have sent a charge of pulverized rock high into the atmosphere. This is the event that blanketed our planet with a layer of fine dust, minute tektites, and the iridium anomaly. A globe-encircling cloud of fine dust would have had several harrowing effects. Scattering sunlight, the dust cloud would have darkened the sky and lowered temperatures everywhere. Some indication of the effect of high-flying dust can be inferred from the recent eruption of Mt. Pinatubo in the Philippines, which caused the average temperature in the northern hemisphere to drop by about 1°C (1.8°F) for most of a year. Though we will never know exactly how long the end-Cretaceous dust-cloud persisted, it changed the history of life.

More important, the asteroid hypothesis helps us understand why some creatures survived and some didn't. The hypothesis implies that effects on animal life in the ocean would be especially severe. Minute phytoplankton near the ocean surface carry on photosynthesis, and, if this source of energy is diminished, life in the entire water-column collapses. Unlike forest and swamp, there are no thick layers of decaying vegetation at the bottom of the sea. Analyses of geological cores off the shore of New Jersey have found debris from the end-Cretaceous impact above deposits containing the microscopic shells of many minute marine organisms. Above the debris layer, nearly all the microscopic species are gone, topped by thousands of years of deposition before the fauna is fully restored. These cores, more than a thousand miles from the impact site, bear witness to the near-instantaneous death of marine planktonic creatures.

On land, most insect groups and some mammal lineages managed to survive the extinction event. But the dinosaurs were wiped out. In contrast, those larger animals that could endure longer time periods without food, such as crocodiles, turtles, snakes, and lizards, survived. (Here is strong evidence that dinosaurs were dynamic warmer-blooded creatures, needing lots of food on a regular schedule.) Several distinctive bird lineages died out, but many survived to become today's bird fauna. More significantly, since the impact occurred in the Northern Hemisphere, it may be that extinction was less severe in the Southern Hemisphere. Land plants show some losses after the impact, but spore and pollen diversity soon recovered. Overall, there is little doubt: though dinosaur numbers may have been declining in the face of volcanism in the late Cretaceous, the final blow was a catastrophic impact of extraterrestrial origin.

BIODIVERSITY EXPANDS WITH EVOLUTIONARY TIME

Despite extinctions large and small, the history of life on planet Earth seems to be one of expanding numbers and increasing complexity. For larger life forms, it all began rather tentatively with the enigmatic Ediacarans. These leaf-like and pancake-like beings are the first larger biological organisms found in the fossil record, 560 mya. They were followed by a short period of small shelly fossils, together with a few trails and burrows. Then suddenly, around 540 mya, a variety of larger fossil forms are evident. The Cambrian explosion had begun, transforming the oceans into a zoo of differing creatures. Animal families diversified further, and their numbers continued to rise during the Ordovician. A severe extinction ended the Ordovician and ushered in the Silurian (443 mya). Moderate ups-and-downs in family diversity continued over time, but then came the catastrophic Permo-Triassic extinction (250 mya).[25] Nevertheless, plants and animals recovered and began diversifying once again. By around 150 mya, biodiversity had surpassed the numbers of species alive before the great extinction event.

Grand extinctions in the history of life allow geologists to sequentially divide the fossil record. In fact, the two largest extinctions allow us to divide animal history into three grand epochs. The first is called the **Paleozoic** (560 to 250 mya) and was terminated by the end-Permian extinction. The **Mesozoic** (250 to 65 mya) was the second act, terminated by the end-Cretaceous extinction. The third epoch, the **Cenozoic**, began with the end-Cretaceous extinction and is arbitrarily terminated at the end of the last ice age.[26] Following the three eons described at the beginning of this chapter, this last 560 million year eon is called the **Phanerozoic**.

Major extinctions have done more than punctuate the history of life. Mammals triumphed only after the dinosaurs departed, and they dominate land surfaces today. The destruction of entrenched faunas created new ecological opportunities. Niles Eldredge has argued that extinctions played a significant role in the progressive evolution of animal life, as old faunas were devastated and replaced by new faunas.[27] Plant history, however, is very different, with major extinctions playing a lesser role. Surviving in the soil as seeds or spores, plant lineages have managed to make it through both major and minor calamities. The grand extinctions did cause extensive ecological upheaval; millions of years were necessary before ecological recovery was complete after the Permian period.[28] However, nothing in the history of plants is similar to the radiation of mammals once the dinosaurs were gone.[29] Surely this is due to more intense competition and predation among animals: a world "red in tooth and claw"!

In contrast, the botanical trajectory across time is marked by new structural innovations. A vascular system allowed plants to become erect; a tubular cambium produced wood and tall trees; pollen allowed seed plants to reproduce in drier habitats. Finally, flowering plants gave rise to a more diverse, dynamic, and nutritious vegetation. Some years ago, paleobotanist Norman Hughes estimated the increasing numbers of vascular plant species over the last 300 million years.[30] Vascular plants, you'll recall, include the flowering plants, conifers, cycads, ferns, and fern allies. These are the major players in building three-dimensional terrestrial vegetation. Hughes suggested that there were only about 500 species of vascular plants worldwide during the Carboniferous period, around 300 mya. By 150 mya, his estimate rose to 3,000 species. At the end of the Cretaceous, 65 mya, Hughes' estimate rises to 25,000 species. Today's total of vascular plants numbers more than 275,000 species. Though Hughes' estimates were highly speculative, they imply an extraordinary increase in land plant diversity over the last 300 million years. More important, Hughes' numbers suggest an expansion from 3,000 to perhaps 300,000 species over just the last 150 million years! While ferns and mosses also expanded their numbers over the last 150 million years, it is the flowering plants that account for much of the recent escalation.[31] And with more flowering plants came more beetles, more ants, more birds, more mammals.

Though nowhere near as numerous as flowering plants, birds and mammals have fostered a more subtle kind of amplification in species numbers, and they did this thanks to what biologists call **parental investment**. Birds and mammals are complex, intelligent, high-energy creatures. They are expensive, both to produce and to maintain. Young birds and baby mammals cannot be left to fend for themselves; they need high-quality food in regular servings. That's what parenting is all about. Unlike most amphibians and reptiles, neither birds nor mammals can let their newborn offspring fend for themselves. In fact, diversification of passerine birds over the last 25 million years may have been spurred by their ability to build sturdy nests in secluded sites, protecting their young in this way.[32] By having to provision their young, both birds and mammals have expanded biodiversity.

Both the young eagle and the young lion will become active hunters only after coming close to their adult size and weight. Little lions do not chase rabbits; young eagles cannot leave the nest until their wings are fully formed. What this means is that both lion and eagle become active players in their ecosystem only when they can play the same roles that their parents do. Since the lion pride provides for the growing cubs, young lions will not begin to hunt until nearly fully grown. The same holds true for virtually all birds and mammals. Compare this with a tyrannosaur. Little tyrannosaurs probably had to hunt little prey, gradually growing to their awesome adulthood over more than twenty years. This meant that tyrannosaurs hunted a wide variety of prey over their life span, and there simply wasn't as much *room in the landscape* for other, smaller, predators. Paleontological evidence indicates that ecosystems dominated by the dinosaurs were relatively poor in animal species. Compare that scenario with Africa's savanna today, where many predators all share the same landscape. Thanks to intense parental investment, bird and mammal species are neatly restricted to their ecological roles, allowing for finer niche division among a variety of similar species. Prolonged parenting in birds and mammals has grandly enhanced animal diversity.

An overview of our planet's extraordinary history makes something very clear: Terrestrial plant and animal life has increased in both complexity and species numbers over time. Putting all the data together leads to a simple but profound conclusion—contemporary biodiversity is higher than it has ever been before.[33] Unfortunately, since becoming fossilized is more likely within the sea, a huge majority of fossil data is that of marine animals, not life that lived on land. Marine fossils also record increasing family and generic diversity over time.[34] Though earlier estimates of expanding marine diversity have recently been scaled back, a clear evolutionary trend appears to be evident.[35] An analysis focused on marine ecosystems themselves concluded that these have become more complex over the last 250 million years.[36] Contradicting these studies, another analysis suggests that marine fossil diversity has not increased greatly over the long haul; the increase in species is simply a reflection of more abundant fossil exposures over more recent geological history.[37] Apparently, the question of increasing marine diversity has not been resolved. Perhaps all the "parking spaces" were taken long ago and the only thing of note is having old species replaced by new species?[38] Such a scenario may be reasonable for marine life, but it stands in utter contrast to what's been happening on land. The terrestrial record is incontrovertible: species richness has increased enormously over the last 400 million years.[39]

Here is one of history's great enigmas! Why, after each and every extinction event, did animal and plant numbers not only regain their previous numbers but add even more? Why have land plants become both more complex in structure and more numerous in species over the last 400 million years? Next, let's examine the forces responsible for greater diversity and complexity in life over time.

Chapter 8

A WORLD OF
EVER-INCREASING COMPLEXITY

*Rather, evolution is a general-purpose and highly powerful recipe
for finding innovative solutions to complex problems. It is a learning
algorithm that adapts to changing environments and accumulates
knowledge over time. It is the formula responsible for all the order,
complexity and diversity of the natural world.*

—Eric Beinhocker[1]

We have just reviewed, however superficially, the rich diversity of Earth-bound life and its long history. But how did all this come about? Why should our lovely planet have so many species of plants and animals? Why are there lush rain forests and flower-filled prairies? As we've just seen, the fossil record makes evident how this glorious diversity has developed and enriched itself over time. Our precious planet has provided a comfortable and stable foundation for the advance of biodiversity over time. It was Copernicus who explained how the Earth, spinning at a tilted angle as it circles the Sun, provides us with the seasons. Each year, this same tilt has the Sun moving north and south, sweeping monsoonal rains back and forth across the tropics, enhancing tropical biodiversity. Today, we understand how plate tectonics has forced mountains high into the sky; another boost for the world's biodiversity. An extensive fossil record makes clear that larger marine life first appeared four billion years *after* the solar system's formation. Terrestrial vegetation and land animals were more recent innovations.

Life's grand proliferation over time has been propelled by two principal factors. Perhaps the most common is that of simple speciation, as lineages split to colonize ever more habitats. Competition, both within and between species, has driven this proliferation. Evolution, however, has been marked by yet another powerful factor: *increasing complexity over time.* Here, simpler forms of organization have given rise to more elaborate levels of organization. Beetles are a lot more complex than earthworms, both in their architecture and in their life trajectory. Jellyfish are even simpler than earthworms, but far more complex than single-celled life. Though instances of escalating structural complexity have been quite rare in the history of life, their significance has been grandly amplified with time.

Endosymbiosis helped build the larger and more versatile eukaryotic cell. Powered by their mitochondria, these cells were able to carry more information and—with time—fashion larger organisms. Once multicellular animals rolled their embryonic cells into a hollow ball and invaginated, they created two layers of cells, which then exploded into a riot of designs. Likewise, and together with a new photosynthetic partner, plant cells began building a world of more nutritious greenery. Slowly, escalations of design, coupled with the simple drive of diversifying speciation, raised the biosphere to new levels of interaction and ever more elaborate ways of living.

> *Evolution manifests a number of tendencies but the most visible of these trends is the long-term move toward complexity.*
> —Kevin Kelly[2]

Unfortunately, the concept of complexity eludes simple definition. Though dictionary synonyms include "consisting of parts" and "complicated," science can't seem to agree on how to quantify complexity. Is it a measure of the energy needed to put a complex object together? Might it be *algorithmic information content* or *internal levels of hierarchy*? Must complexity have internal nodes for relaying information? Such efforts quickly descend into a philosophical muddle. (Using many examples from the natural world, Melanie Mitchell clarifies these challenges in her book, *Complexity: A Guided Tour.*)

Mathematicians have also examined complexity. But, as Steven Strogatz points out, "Complexity theory taught us that many simple units interacting according to simple rules could generate unexpected order. But where complexity theory has largely failed is in explaining where that order comes from, in a deeply mathematical sense, and in tying the theory to real phenomena in a convincing way. For these reasons it has had little impact on the thinking of most mathematicians and scientists."[3]

Despite these problems, I think we can move right along using **complexity** as common usage defines it. Forests are complex because they are home to many living things; we are complex because we have many interdependent parts built by trillions of individual cells. But what is it that has propelled seemingly universal trends of increasing complexity in so many aspects of the living world? Over time, molecular, organismic, and environmental systems have all become more complex. When we discussed speciation, we mentioned a number of factors that result in the creation of increasing numbers of plant and animal species. This is a kind of multiplication; but why does evolution appear progressive as well? Daniel McShea and Robert Brandon answer this question in bold fashion by describing a "New Biological Law." They claim that, in the absence of calamity or strong countervailing selection, biological systems naturally become more diverse and more complex. Their **Zero Force Evolutionary Law** (ZFEL) declares that, over time, diversity and complexity will increase, on average.[4] Reproductive systems that cannot reproduce themselves with 100 percent accuracy will necessarily produce variation, and this variation will inevitably increase diversity. While Darwin's **natural selection** requires three components—reproduction, variation, and selection—the Zero Force Law requires only two—reproduction and variation. And since reproduction, at whatever level, cannot achieve 100 percent accuracy (thanks to Murphy's Law), increasing diversity and complexity are inevitable. Only strong stabilizing selection can keep living things from changing over time (stasis). Thus, today's world is far more complex than in times long past. Though a huge majority of mutations are deleterious, selection sweeps them away—even as a tendency to diversify (ZFEL) continues undiminished.

But what keeps the entire system moving forward? The answer to this question is simple: **energy.** In fact, *all living things* move energy around within the cell in the same way. Adenosine triphosphate (ATP) acts like a little battery; by losing its third phosphate group, ATP donates energy to wherever it is needed, becoming adenosine diphosphate (ADP). In turn, ADP uses energy-yielding reactions within the cell to re-acquire the third phosphate group, becoming ATP once more, ready to donate energy with the loss of its terminal phosphate group. But what are the sources of this living energy?

THE FIRST GREEN REVOLUTION

Energy is required to do work, whether moving a boulder or digesting your dinner. Because of this, energy is central to the survival of all living things. Here on planet Earth, we have only two primary sources of energy. Rocks deep beneath our feet still retain heat from Earth's early formation. In addition, unstable atoms within the Earth's core continue to break down, releasing nuclear energy. Together, these sources keep the Earth's center molten, magma extruding at deep-sea fissures, and volcanoes erupting. The strange biota of deep ocean vents are powered by the Earth's own deep energy. There, bacteria use chemical reactions to power their life activities and, in turn, support strange creatures in dark ecosystems.

Fortunately, planet Earth has a second and more accessible source of energy: **sunlight**! In the past, a few critics derided Darwin's evolutionary scenario because, they claimed, evolutionary progress contradicted the second law of thermodynamics. The second law tells us that everything in the universe *runs down* or moves to a state of lower energy and diminished organization. Evolution on planet Earth has done exactly the opposite! Not to worry: the **second law** applies to closed systems, and planet Earth is not a closed system—as anyone who has suffered a sunburn knows! By circling the Sun in the "Goldilocks orbit," the Earth and its atmosphere maintain just those temperatures needed for keeping water wet. Clearly, the second law holds true: the Sun is "running down" on an enormous scale. Intercepting less than a billionth of that outward flowing radiation, our days are sunny, allowing photosynthesis to power much of the living enterprise. Though a number of bacteria can make a decent living at the edge of lightless sea-floor vents, or by using the energy of chemical degradation, all the rest of us are powered by our star.

Once living things learned how to capture and utilize the energy of sunlight, evolution began running uphill. One of the most crucial advances in the history of life, water-splitting photosynthesis provided hydrogen for building carbohydrates.[5] No simple task, pulling apart the water molecule required a suite of complex molecules—acting in exquisite unison. At a minimum, ten photons of light are necessary to 'fix' one molecule of carbon dioxide in the business of building carbohydrates. Simply stated: **photosynthesis** transforms the physical energy of solar radiation into the chemical energy of food. Understanding photosynthesis required many years of research. Paraphrasing one of its early investigators: **radiation physics** was needed to understand the light; **solid state physics** detailed the light-capturing process; **physical chemistry** helped clarify initial oxidation; **biophysics** explained the electron-transport system; **biochemistry** helped us understand carbon dioxide fixation; **plant physiology** showed how the biochemistry was regulated; while **botany** situates these processes within the plant, and **ecology** makes clear how photosynthesis powers the environment.[6]

Fossil evidence, based largely on isotopic changes in ancient sediments, indicates that the Cyanobacteria had developed water-splitting photosynthesis at least 2.7 billion years ago.[7] That was the good news: using the energy of sunlight and hydrogen from water to build energy-rich carbohydrates. The bad news was that highly reactive oxygen was being loosed upon the Earth. A new and threatening pollutant, free oxygen began to change the world. Because of its reactivity, oxygen unites with many minerals, or burns, and quickly leaves the atmosphere. However, with **oxygenic photosynthesis** now in play, bacteria were pumping oxygen continuously into the atmosphere, millennium after millennium. Though still a dangerous poison for some bacteria living today, in those early times a few bacteria devised a way to *use oxygen!* They reorganized their energy-acquiring metabolism to "burn" free oxygen. We call this oxygen-utilizing metabolism **respiration**. Oxygen-devouring respiration extracts ten times more energy from the breakdown of carbohydrates than does fermentation! Bacteria with this new metabolic system had a clear advantage. Simply stated, photosynthesis pulls apart the water molecule and respiration puts it back together again—a perfectly balanced cycle kept running by the power of our Sun.

INCREASING COMPLEXITY BY ADDITION

The emergence of oxygen-utilizing respiration laid the groundwork for another major advance in the history of life. Turns out, one of these respiring (oxygen-burning) bacteria became part of the eukaryotic cell. We call these organelles **mitochondria**. With oxygen more widely available for respiration, and powered by their mitochondrial fuel-cells, eukaryotes became the platform for further advances in biological complexity. Delivering hydrogen (from the breakdown of sugars) to oxygen, mitochondria gave complex eukaryotic cells the power to handle more elaborate genetic instructions and, eventually, produce more complex organisms. What respiration accomplished was to utilize a newly available resource—free oxygen—as a means of acquiring more energy.

Today, the division between the bacterial world and those organisms with larger nucleated (eukaryotic) cells is clearly the most significant division on the tree of life. Most bacteria lack sufficient volume to house a nucleus or other organelles. By sequestering DNA within the nucleus, fundamental genetic information was protected from the creative and destructive dynamics of metabolism. With more internal complexity and greater information-carrying capacity, the larger eukaryotic cell proved to be a singular advance in the history of life.

Just as the incorporation of the mitochondrion within the eukaryotic cell was a major step forward in the history of life, another bacterial endosymbiosis empowered yet another eukaryotic lineage. By making

Cyanobacteria partners within their own cell walls, **green algae** became photosynthesizers themselves. As with mitochondria, these photosynthetic symbionts became essential organelles within plant cells; we call them **chloroplasts.** Using the same chemistry as their bacterial antecedents, chloroplasts absorb red and blue light, empowering themselves and those who eat them. By reflecting and transmitting the green light they're not absorbing, plants have turned our landscapes green.

These grand stages in the history of life, however, posed a very serious question for earlier biologists. How could minor and random mutations have fashioned so great an increase in living complexity? Clearly, addition is something more elaborate, and this is why the concept of **endosymbiosis** proved so satisfying. Discovering that both mitochondria and chloroplasts still carry a few of their own genes confirmed the hypothesis; these organelles had once been bacterial-grade organisms themselves. This was not a case of natural selection choosing among slightly differing base-pair sequences or new mutations. Instead, two highly structured entities came together to create a new and significantly more complex cell! Thanks to this very special kind of amalgamation, both respiring eukaryotic cells and photosynthesizing algae became major advancements in the history of life.

Mitochondria energize all larger, more complex organisms, from amoebas and fungi to plants and animals. And again, greater energy, supplied by oxygen-consuming respiration, allowed the eukaryotic cell to grow larger, house more chromosomes, store more information, and power the division and multiplication of a cell both larger and more elaborate than the bacterial cell. This innovation may have occurred as early as two billion years ago, but another billion years would pass before more elaborate multicellular plants and animals came upon the scene. And while free oxygen was being produced continually by Cyanobacteria in these earlier times, its presence in the atmosphere was minimal. Not until about 600 mya did oxygen pressure begin to rise, followed by the Cambrian Explosion of larger animals.

ANIMALS EVOLVE GREATER COMPLEXITY

Becoming a larger and more complex animal entails many costs. For starters, more energy is required when building a larger being. Subtle protocols are necessary to have hundreds of cells multiply and differentiate to join into a single coherent whole. A *spirit of cooperation* must be maintained as ever-greater numbers of cells unite to form the larger organism. Each cell must confine its own destiny to becoming a functional member of the larger individual. Differentiated cells within the larger organism perform specialized functions, providing the benefits of a *division of labor*. We humans have specialized cells to form our skin, our intestines, our muscles, and our brains. Though becoming bigger and more complex entailed many costs, the benefits have been astounding: from gelatinous jellyfish to many-legged millipedes, exquisite butterflies, and lumbering elephants. Larger animals garner more resources, cover more territory, and may reproduce more successfully. Unfortunately, greater organic complexity is fragile, ending in death and dissolution.

There's no avoiding the obvious: over time complex things break down. We are born, we grow, we reach maturity and reproduce, but then we decline and perish, to be superseded by the generations we helped bring forth. There comes a time when fixing an old machine becomes too costly, and it has to be abandoned. Larger, complex creatures eventually die. Not so bacteria or single-celled microbes! Their simplicity allows them to continue splitting in two for as long as their luck holds out. They are, in effect, immortal. While the human genome can be thought of as being immortal, changing only slowly over thousands of generations, we ourselves are mortal. Nevertheless, multicellular life has transformed the planet in ways the bacteria could not.

How did single cells paste themselves together, learn to communicate with each other, abandon their own agendas, and become organized into larger, more complex beings made up of thousands to trillions of cells? Second only to the enigma of life's origin, this question remains a major challenge for developmental biology. Free-living cells have a fundamental program: *divide and multiply!* Make more cells! But cells living within a complex organism cannot divide or multiply uncontrolled. To do so will destroy the organism itself. We know this only too well, and describe it with a single frightening diagnosis: **cancer.** When cells begin to multiply unconstrained within our bodies—no longer playing by the rules—we are in mortal danger. Think about this: each one of us is made up of *trillions of cells working together to form a single integrated individual.* Amazing! How do these many cells manage to work together, and, how were they assembled? These questions fall into two categories. The first is how did each one of us become the hugely complex individual we are? That's a developmental question, going back to our early embryonic origins. But here we are concerned with a more general question: How did the ability of complex animals to fashion themselves develop over time?

To build a larger organism, cells have to be able to both cohere together and communicate with each other. Once they were united, how did a system of constraints—a system that is basic to the functioning of all complex living things—develop? Only a delicately balanced system of intercellular communication can orchestrate coordinated development. Tissues need to develop where they are supposed to develop—and nowhere else! In developing animals, some cells must get out of the way to allow others to expand. This requires self-destruction (programmed cell death, or apoptosis). All the while, cells must duplicate in exquisite synchrony. We really do not know how our left and right arms end up so perfectly matched. Fortunately, most bilateral animals do have their right and left sides nicely symmetrical, so they can swim, run, or fly effectively. And finally, once the young animal has achieved a mature size, it must cease growing and begin reproducing.

Certainly, biological development must be fine-tuned to produce and maintain symmetrical forms of proper size, whether beetle, lizard, or whale. A closely cadenced harmony of genetic instructions, developmental protocols, and self-regulating networks has fashioned the world's larger life forms. Organisms that failed the challenges of symmetry and cellular harmony are long gone. Here is an evolutionary story that began over 500 million years ago and has been transformed into a developmental program repeated in each and every one of us. Recall that every leafy plant, feathery bird, and furry mammal arises from a single fertilized egg cell. Indeed, our development does recapitulate—in a general way—the major stages of our long evolutionary history.

For animals, forming new layers early in the life of the embryo was a key innovation. After a number of cell divisions, the fertilized egg transforms itself into a hollow sphere only one cell thick. Then, one side bends inward and invaginates to form a two-layered sphere with a little hole outside. The future of these two layers, inside and outside, are very different. Each will contribute different tissues to the developing animal. The little hole that's left outside can become a mouth (in snails, squids, crabs, and insects) or an anus (in starfish and vertebrates). Each of these two very different groups of animals then proceeds to develop a second little hole, providing another opening, front or rear. Having an input-hole at one end and an exit-hole at the other provided an efficient one-way digestive system! (Flatworms, sea anemones, and jellyfish have only one opening and represent an earlier stage of animal evolution.)

Both the development of a human baby from simple embryonic beginnings and the elaboration of a mature butterfly within its pupal skin are awesome transformations. The fact that such developmental trajectories have produced millions and millions of humans and even more beetles, should not diminish our astonishment. As Neil Shubin points out, "Like a concerto composed of individual notes played by many instruments, our bodies are a composition of individual genes turning on and off inside each cell during our development."[8] The discovery that plants, animals, and humans use similar gene systems to guide their development is evidence for common origins very long ago.

What are the essential components in this story of increasing complexity? Genes are the starting template; they fashion a wide variety of proteins that do the work of the cell. Regulatory DNA activates genes when and where they are needed. Changes in developmental timing are a major source of evolutionary novelty. All the while, within-cell metabolic networks maintain stability (homeostasis), essential to keeping life processes running properly. In turn, cell-to-cell interconnectivity and communication build and maintain the larger organism. Cells generate ordered complexity by responding to subtle inputs from adjoining cells; they work together in a matrix of connectivity, producing larger tissues and organs. These, in turn, must be coordinated by a complex nervous system, keeping the entire organism on a proper course. Animals and plants use a variety of non-linear processes to form interactive networks that result in *emergent self-organizing systems* of great structural complexity. Each of us has developed and grown in exactly this way. But the more extraordinary claim is that natural processes, slowly fashioned over these last three billion years, have resulted in the biodiversity we see today.[9]

ELABORATING ANIMAL COMPLEXITY

One of the unusual revelations of the late twentieth century was that all animals share many of the same developmental genes. An eye-gene from a squid, placed on the embryo of a fruit fly, produced an eye at that location: a fruit fly eye, not a squid eye! Clearly, the "put-an-eye-here-gene" is the same for the squid and the fruit fly. Similar results were obtained transplanting mouse-eye genes. Again, the fly developed a fly eye at the site of the transferred mouse gene. These very different organisms, with very different kinds of eyes, have been using similar genes to build their very different eyes. No one had thought that the camera-like eye of squid or mouse had a common origin with the multi-faceted eye of an insect. And, in fact, they don't. But different lineages did use the same genetic tools in fashioning their different eyes over evolutionary time. Moreover, all animals use light-sensitive opsin molecules in the business of sensing light. We humans have three genes for making three slightly different opsins, allowing us to see the world in a great variety of colors. Apparently, where one gene would produce one opsin, a duplicated gene could produce a slightly different opsin, and the beginnings of color vision. Where one developmental gene might initiate a single lobe-like appendage on a velvet worm, additional—accidently duplicated—genes could create the several-jointed leg of an insect. Genetic analysis has revealed how each animal lineage came to have its particular *Bauplan*.[10]

Not that long ago we thought that humans had at least 100,000 genes. Why not? We are hugely complex beings and very, very smart! But after further study, it turns out that we've got about 24,000 protein-coding genes.[11] Worse yet, the species closest to us in both form and behavior—the chimpanzee—shares between 94 and 98 percent of our genes (depending on how you score the genome). Unfortunately, these enumerations focused on protein-coding genes, and that's only a part of the hereditary story.

Let's change our focus for a minute, and look at plants. It turns out that our favorite laboratory plant, the thale cress (*Arabidopsis thaliana*), has about 24,000 genes! At the time of discovery, this was a huge surprise. Our other laboratory favorite, the fruit fly (*Drosophila melanogaster*) was estimated to have only about 13,600 protein-coding genes. Think about this! The weedy thale cress grows to little more than a foot in height. Then it just sits on the ground capturing sunlight to make flowers and produce seeds. The fruit fly, in contrast, has a full four-stage life cycle, can fly hither and yon, and has males that not only search for females but can dance little dances to entice the ladies. One would think something as complex as a fruit fly ought to have lots more genes than a static weed. This is significant. The few genes that differentiate *Homo sapiens* from *Pan troglodytes*, or the numbers of genes in the thale cress are telling us something.

Obviously, the number of protein-coding-genes is not the whole story. This is especially true in the plant-animal comparison. Cottonwood trees have recently been shown to have around 44,000 genes! Further genetic studies have indicated that many parts of an animal genome determine RNA sequences that are important in controlling the timing of cellular and developmental functions. In fact, RNA processing and modification are more versatile in animals and may explain why fruit flies have "fewer genes" than the little plant. This may also explain why we humans differ so profoundly from chimps, despite sharing so many genes. There is an intricate developmental trajectory that takes us from our genetic blueprint (our **genotype**) to what we end up looking like (our **phenotype**).

Problem was, geneticists began with a far too simplistic concept of "the **gene**." Protein-coding genes make up less than 5 percent of our genome; the rest was called "**junk DNA**." But if it was junk, why hadn't natural selection gotten rid of it? Turns out that much of our genome maintains the organization of the chromosome, regulates gene expression, codes for RNA regulation of cell activities, and does carry some "junk"—bits of viruses and "fossil genes'" that are no longer functional. There's lots more to do than code for proteins. And here is a more significant point: larger amounts of "junk DNA" are correlated with the increasing complexity of the animals that bear them.[12] Complex life depends on many different gene networks, precisely timed developmental cues, flexible physiology, dynamic morphogenetic fields, and lots more.[13]

Our early development (early ontogeny) indicates how developmental protocols have changed over evolutionary time. We humans build and replace two rather different kinds of kidneys during our early development. As you might imagine, a fish-like kidney forms first, but is resorbed. A kidney rather like that of reptiles forms next, and this too is abandoned. Finally, we end up with a mammalian kidney that serves us for the rest of our lives. Such embryological sequences prompted Ernst Haeckel, evolution's early proponent in Germany, to declare that "ontogeny recapitulates phylogeny." Early embryogeny and development provide further evidence for our long evolutionary history. Blood vessels go down, around, and up again in the region of our throat. If you've got a Y-chromosome, testicles formed within your body, then exited through the abdominal wall to keep themselves cool (leaving a weak spot and making a hernia more likely). Such observations are testimony for our long evolutionary history (they do not support a "six-day creation 10,000 years ago").

"The evolution of complex creatures is no mere epiphenomenon; it is one of the marvels of the universe" claims Wallace Arthur. "From no head to rudimentary head to well-developed head to sophisticated head is definitely a series of steps up the ladder of complexity."[14] Just take a close look at a bug's head; check out the fancy mouthparts, big eyes, and slender feelers, all exquisite and *purposeful* parts. Though bacteria and jellyfish have remained constrained within their particular levels of organization over hundreds of millions of years, other lineages explored new possibilities. And because life had arisen within a world of water, living beyond the liquid realm necessitated new and extraordinary innovations.

THE SECOND GREEN REVOLUTION: A TERRESTRIAL FLORA

In the introduction we noted that today's biodiversity—in the sense of larger life forms—manifests itself most grandly on the land. However, moving from an aquatic lifestyle to survival on dry land was a major initiative, and especially difficult for plants. The key innovation may have been elaborating the plant's diploid state to become a more important part of the life cycle. The sperm of algae ancestral to land plants (Charophyta) must swim to the egg cell to form a zygote with two sets of chromosomes (the diploid condition). This produces a short-lived plantlet, which soon undergoes meiosis, building another generation of haploid algae that will produce new sperm and egg cells. Here, the diploid stage is short, and the entire lifespan is underwater. How might these aquatic algae have changed to live at least a part of their life span on dry land?

Land plant genomes possess two sets of developmental KNOX genes, while their ancestral algae have only one set! Apparently, the newly duplicated set of KNOX genes allowed the diploid zygote to become a longer-lived diploid plant![15] In effect, early land plants had two prominent life phases: a haploid generation and a diploid generation. The diploid plant—after meiosis—produces haploid spores and a new haploid generation. This haploid generation then produces gametes that will unite and form a new diploid plantlet. Botanists call this the **alternation of generations** and it is an important part of the lives of ferns, mosses, and liverworts. But what made this new two-stage life cycle so important?

Here we have a clear evolutionary advance based on a single duplication within a family of developmental genes: KNOX1 and KNOX2! Imagine the following scenario: our ancestral alga is living in a small pond; it is haploid and produces sex cells (gametes). Swimming sperm fertilize egg cells under water and begin the new diploid generation. Then the dry season begins, our little pond dries up, but the diploid plant now basks in the light of the Sun and lives long enough to undergo meiosis and produce haploid spores! Such spores can be released into thin dry air to be carried far and wide by the wind! In this scenario, water is still necessary for swimming sperm to reach the chemically attractive egg cell and produce a diploid zygote. But, with KNOX2, the diploid stage expanded to become a plant that could survive and disperse its haploid spores over long distances. Here we had a diploid plant able to live on land, and a haploid plant (formed by the spores) still living in water, where sperm could swim to egg cell. Later, after sperm cells "learned" to swim to the egg cell on thin films of water on plant surfaces, the life cycle became completely terrestrial.

Unlike the sudden explosion of larger animal life in the early Cambrian, one hundred million years would pass between the first evidence of land-plant spores and the appearance of tropical forests. Plants faced many challenges before they covered the land in greenery. A waxy surface constrained water loss; stomates regulated gas exchange; roots and plumbing brought water up within the plant, and, later, a tubular meristem would construct tall trees. A bit later, seed plants advanced into drier environments. The terrestrial world was now graced with tall forests and diverse vegetation.

A THIRD GREEN REVOLUTION

Beginning around 130 million years ago, flowering plants fostered yet another grand escalation in terrestrial biodiversity. Allocating less energy to defense and more to growth and reproduction, Angiosperms quickly overtook the other greenery.[16] Colorful flowers and animal pollinators allowed many Angiosperms to reproduce effectively at lower population densities, packing a greater number of plant species into the same biome. More species, greater structural variety, and new flowering and fruiting modalities created environments that were both more elaborate and more nutritious. Nectar-feeders became flower-pollinators. Fruit-eaters became seed-dispersers. Between about 125 and 80 million years ago, the terrestrial fossil record displays a significant pulse of diversification. Given the acronym KTR (K for Cretaceous, T for terrestrial, and R for radiation or revolution), this was a time during which many lineages of both plants and animals expanded their numbers in grand synchrony. In fact, this may have been the moment in time when terrestrial species numbers surged beyond the numbers living within the sea.[17]

Today, plant variety is displayed most grandly within the lowland tropical rain forest; here flowering plants display their greatest structural variety. Trees with wide buttresses bracing tall trunks ascend into the upper canopy, even as thick woody lianas swing down from high above. Broad-leaved aroids adorn the forest floor or clamber up trunks toward brighter light. Palms reach into the lower canopy with broad feathery leaves at the tips of slender, unbranched trunks. Banana-like plants with leaves up to twenty feet tall cluster at stream margins and in clearings. Green foliage varies from simple elliptic leaves to those variously lobed or divided. Surprisingly, one sees few flowers within the rain forest; they've got all year in which to bloom, and most are found in the higher canopy. Greater structural diversity among flowering plants provided more niches for more species. Mosses, little ferns, and small orchids could grow on widespread branches, while primates, birds, and insects fill the canopy with noisy chattering. As we stated earlier: flowering plants have been the primary drivers of expanding terrestrial biodiversity over these last one hundred million years.

CHANCE OR NECESSITY?

Returning to the origin of land plants themselves, their apparent unity-of-origin brings up a troublesome evolutionary question. Stephen Jay Gould phrased it very nicely when he asked, "Would we [humans] appear at all, if we could rewind the tape to an appropriate beginning (say the origin of modern phyla in the Cambrian explosion more than 500 million years ago) and simply let it run again?" Gould thought that the role of accident and **contingency** were so fundamental that most plants and animals would, indeed, be very different "if the tape were run again."[18] Disputing Gould, Simon Conway Morris argued that environmental effects would tend to *select* creatures similar in their adaptations. To support his ideas, Conway Morris cites many cases of **convergence**, where different lineages of plants or animals had done much the same thing—independently—through convergent evolution.[19] Obviously, if you are an animal planning to go somewhere it's good to have your head up front so you can see where you're going, no matter what kind of animal you are. That's a general convergence, but more specific trends are also common. The marsupial mammals of Australia have produced forms that resemble jumping mice, moles, and small dogs, quite similar to those found among placental mammals outside Australia. The tall columnar cacti of American deserts look rather like tall euphorbias in Africa, but they are unrelated, having evolved in widely separated hot dry climates. Surely, both Gould's and Conway Morris's arguments have validity. Both contingency and convergence have been part of life's history.

Unlike the situation among land animals, where a number of unrelated lineages pioneered the land, only a single lineage of plants succeeded in becoming terrestrial. Perhaps, without this "lucky break" Earth might be devoid of complex terrestrial ecosystems. Similarly, if early primates had not pursued insects into the tops of flowering trees, perhaps no similarly versatile—or clever—mammalian lineage could have become human. Both accidental contingencies and selective environmental constraints have determined the course of life's long history—initiating opportunities and then driving lineages down particular paths. But where did new genes and new developmental protocols come from?

MURPHY'S LAW: A WELLSPRING FOR DIVERSITY AND PROGRESS

Tradition has it that there really was a Murphy, an engineer in the United States Air Force during World War II. He proposed a rule for designing control mechanisms, but that rule was soon corrupted and transformed. Murphy's rule had suffered the fate decreed by Murphy's Law. Today, Murphy's Law is simple and universal: **In complex systems, what can go wrong will go wrong!** In the reproduction of living beings, things often go wrong. However, on rare occasions, "errors" can be the basis of adaptive change. Failures will be eliminated: that's what natural selection is all about. Our common intestinal bacterium, the much-studied *Escherichia coli*, suffers mutations with detectable effects all the time. Researchers estimate that each "good" mutation arising in *E. coli* is outnumbered by around 100,000 "bad" ones. Not a good ratio! But clearly, not everything gone wrong is deleterious. In fact, rare mistakes may open the door to new possibilities.

Accidental **gene duplication**, resulting in two genes where one was quite sufficient, provided an opportunity to do new and different things. As we just mentioned, duplication of the KNOX gene may have allowed for the expansion of the diploid generation in early land plants: one of the most important advances in the history of life.[20] Among land vertebrates, one developmental HOX gene gives rise to the upper arm, a second to the lower arm, and a third to the hand. Such development-altering duplications may be even rarer than the 100,000-to-1 mutation ratio reported in *E. coli*. However, over hundreds of millions of years, such rare events have propelled evolution along a trajectory of ever-greater complexity.

Recent work in developmental genetics has made clear the importance of *duplicated* genes. As already noted, we humans have three slightly different light-sensitive opsins, coded during their development by three similar but slightly altered genes. Together, they give us trichromatic color vision. Many other attributes appear to have been elaborated through gene duplication as well. The laboratory mouse has nearly a thousand genes devoted to the business of olfaction—sniffing and smelling. For animals that spend so much time close to the ground, scent provides important information. Marking their territory with scents, they communicate with their neighbors. Elaborating those many genes involved quite a bit of gene duplication, as well as modification of old genes for new tasks. Interestingly, the human genome carries the same olfactory genes as does the mouse. While this may seem like an academic detail, here's the big news: in humans, a large number of these olfactory genes are no longer functional! We no longer possess the olfactory sensitivity our distant ancestors once had, because primates decided to live in treetops. Here scent-marking was unimportant, but seeing in three dimensions and in color was important. Selection worked on our visual talents, even as our ability to smell declined. This loss of ability is now witnessed by the many genes that still sit on our chromosomes but no longer have a function. These "fossil genes" are further evidence of our deep history.

Speaking of adaptation by selection, we must remember that hereditary mechanisms are tightly constrained. Information carried by the DNA is translated into messenger RNA within the nucleus *and then* exported to other parts of the cell to build proteins or regulate reactions. *This process cannot be reversed.* Proteins cannot redirect or reconfigure DNA within the nucleus. No matter what experience the adult suffers, the consequences of that experience does not alter their DNA, safely sequestered in the nucleus. Jean-Baptiste Lamarck was wrong: experiences of a parent's life cannot be added to the information carried by the hereditary material. (Children of the stone mason are not born with stronger arms.) If scientists want to change genetic instructions, they must insert new DNA snippets into a chromosome residing within the nucleus. Today, genetic engineering can do exactly that—using plasmids and other vectors.

Over time, Murphy's Law has provided a continuing source of mistakes, duplications, and rearranged genetic instructions. Winnowed by selection, a very few of these changes have helped organisms modify their development or their abilities and live out their lives in new ways. (Actually, many mutations seem to be "neutral," making no measurable difference.) It is the few good mutations, duplications, transpositions, and *other things gone wrong* that provided continuing variability—to be tested by an unforgiving environment. Sexual reproduction enhances this process by keeping a species' genetic resources constantly in flux, allowing selective processes to continually screen the population. "Cruel and heartless!" you might think, but this process has allowed complex living things to adapt and transform themselves over hundreds of millions of years.

COMPATIBLE NOVELTY, CONSTRAINED DISORDER

Genes do not run our bodies or our lives. Living cells and tissues are incredibly complex, maintained by many interactions. Genes provide the basic instructions and controls, but cell machinery, mostly run by proteins, builds, and runs itself. Bad genes make bad proteins, which can compromise function; really bad genes can stop early development in its tracks, and the story ends there. It is the integrated organism that lives, reproduces, and, ultimately, dies. Each of us began as a plump egg cell inspired by a single sperm. That egg cell was where all our cell membranes began, where all our cytoplasm originated; it housed the mitochondria that would power all our life processes, reproducing themselves with every cell division. Thanks to their complexity, cells are the only source of new cells. All the while, DNA sits protected within the nucleus, duplicating itself with every cell division, and providing the basic instructions as they are called for. More versatile and dynamic, RNA carries instructions beyond the nucleus and participates in many life processes.

Genetic determinism is a concept that explains only a few exceptional traits. In reality, there are a long series of networked interactions that transform an organism's genotype into the characteristics of the mature individual, its phenotype. A multidimensional network of genetic instructions and interacting regulators provides flexibility and redundancy within each cell. Barbara McClintock's "jumping genes" (transposons) can alter the neighborhood of individual genes and change their functioning in that way. Neutral mutations may enhance the elaboration of a particular gene at a later time. A specific enzyme may be coded by different genes with a variety of different configurations, allowing the entire system to adapt and innovate. Within an individual cell, metabolisms are diverse and dynamic. However, the entire system must be robust and maintain effective **homeostasis:** the process of keeping everything in balance within the cell and the organism. We humans are composed of about a trillion eukaryotic cells, arrayed in some 220 different tissue-types. But when we take our temperature, the readings are usually around 98.6 F; this is an amazing feat of self-regulation! While some disorder allows for flexibility and innovation, the stringent control of disorder (homeostasis) has been a necessary requirement throughout the history of life.[21]

Key innovations, major breakthroughs, and new levels of complexity are all convenient ways to mark major steps in the history of life. Special attributes help us demark major lineages. Cladistic systematics defines lineages by their unique traits. Mammals, in contrast to other land vertebrates, have hair, nurse their young, and possess three little bones within their inner ears. Similarly, flowering plants have seeds enclosed within a protective ovary and use double fertilization to initiate seed development. These important and unique characteristics distinguish mammals and flowering plants from related lineages. However, it seems unlikely that such key innovations sprang into being suddenly. The idea of a single major mutational change (a macromutation, or a "Hopeful Monster") suddenly ushering in new possibilities may satisfy simplistic views, but is utterly unlikely. Whether for a bacterium, a eukaryotic cell, or a multicellular organism, innovations must come in small steps that do not disturb dynamic balance within the organism. Whether at the level of the cell, the tissue, or the individual, dynamic integration between all the relevant processes must be maintained. A major new mutation in one aspect of the cell's functioning cannot disrupt other vital processes. The dynamic network can be tweaked and shifted, but if other elements in the network are adversely affected the new mutation cannot spread through further generations. Selection acts upon the entire organism; if the new mutation compromises the well-being of its host, it has little likelihood of being incorporated into that species' future. To be maintained, "selfish genes" cannot impair the lives of those who carry them. The integrated functioning of the **phenotype**—the actual living being that the **genotype** produces—is paramount. And this means that genetic changes are most often minor, allowing all the players in the network to adapt and adjust throughout the developing organism.[22]

One of life's grandest *progressive steps* took place when one lineage of fish began to adapt to a new and challenging environment. The fish-to-amphibian transition took millions of years and required many subtle changes. As we mentioned in the last chapter, this advance began with lobe-finned fish having to negotiate obstructions in log-jammed streams and estuaries. Lobed-at-their-base, these fins had more flexibility than fins firmly planted on the body, allowing this particular lineage of fish to begin a new trajectory. Lobed fins soon developed something new—elbows—for greater flexibility. Pushing themselves through obstructions gave rise to further innovations: shoulders to attach muscles, and a neck to give the head movement. Snapping-up prey along stream sides fed these "evolving fish." This scenario required more flexible forelimbs, a shoulder girdle to give the head and neck more independence, and digits for better traction. Quick-attack made greater demands on the central nervous system, vascular system, and musculature of the front limbs. Becoming better aerial hunters required better vision, olfaction, and hearing. Slowly a fish-like animal adapted to the land and transformed itself into a four-limbed tetrapod. A constellation of anatomical and physiological characteristics were *changing together over evolutionary time.* These hungry fish were not "evolving to live on land," they were stalking tasty invertebrates along the shoreline.[23]

IT'S A WAR OUT THERE: THE RED QUEEN'S ADMONITION

Biologists often talk about **fitness.** "Survival of the fittest" became a catch-phrase soon after Darwin published his bold new view of how life had elaborated itself over time. And, logically, fitness came to be equated with reproductive success; the more of your genes that become part of the succeeding generation, the *more fit* you have been. Critics complained that this made evolution-by-natural-selection circular! Fitness defined evolutionary success, and evolutionary success was defined by fitness. What these dimwits failed to comprehend was that fitness—measured as reproductive success—was a *process over time.*[24] Operating over time and across generations, advancing fitness couldn't possibly be circular. Biologists have continued working with the idea of fitness, mostly from the point of view of the individual organism and its reproductive success. However, with a little bit of imagination, there's another way of looking at fitness.

Leigh Van Valen of the University of Chicago approached the notion of fitness from a new perspective by thinking of fitness in terms of the overall species, rather than any individual within that species. He suggested that **fitness** might be a measure of an entire species' use of energy in the ecosystem in which it is living. And, he reasoned, if a species is increasing its overall fitness over time, it must be taking a greater bite out of the energy available in that ecosystem. After all, the amount of energy in any ecosystem is tightly constrained, ultimately dependent on solar energy. From this premise, Van Valen argued that as an individual species increases its fitness it must affect all other organisms in the system *negatively!* And, just as the Red Queen advised Alice in the story by Lewis Carroll, Van Valen suggested that each species finds itself in a world where "It takes all the running you can do, to keep in the same place!"[25] Surviving in dynamic communities demands constant adaptation, not only to the physical aspects of the community but also to other ever-changing biological members of the community.

The **Red Queen hypothesis** highlights the competitive interactions between species living in the same environment. Other paleontologists studying extinctions over time had concluded that the random trials and tribulations of changing climates and other minor calamities were paramount. Focusing on unpredictable changes in the *physical aspects* of the environment, this has been called the **Field of Bullets hypothesis**.[26] Here, a sudden physical event can terminate your existence just as effectively as a bullet. Actually, both the "Red Queen" and the "Field of Bullets" are reasonable scenarios; they are not mutually exclusive.

The pressure to *keep running* is not just in response to our parasites, pathogens, and things trying to make a meal out of us. We must compete with others of our own kind for nourishment, territory, and mates. Competition is an overall dynamic, sculpting many aspects of the ecosystem. Unless you find a special corner in which to hide, you better be able to keep up with the crowd. One way animals can "hide" is through camouflage, making it difficult for predators to find their prey.[27] Mimicry allows a harmless insect to resemble a stinging wasp, or a moth to fold its wings and "become" a dead leaf. Both camouflage and mimicry are graphic evidence for a perilous world.

Over time and with new "inventions," many lineages were able to improve their competitive standing. Vascular tissue in land plants allowed them to grow taller and shade smaller non-vascular plants. Pollen allowed seed plants to expand into drier climes. Recall the lichen symbiosis, which allowed these "composite organisms" to thrive in some of the world's most severe environments.

Pest pressure and incessant competition have populated the harshest environments, and the reason is easy to understand. In a dangerous world, peripheral environments may be physically more demanding, but less challenging in terms of disease and competition. John Bonner called this the **pioneering effect** and argued that this leads to the invasion of increasingly stressful habitats.[28] Bonner's scenario claims that competition, together with disease and parasitism, have been the driving force populating some of the world's most uncomfortable habitats. Recall the Archaea; their ability to live in boiling water, high acidity, or high alkalinity may have taken hundreds of millions of years to develop, and they now have these nasty environments mostly to themselves.

Bonner's suggestion reveals a serious flaw in discussions regarding the possibilities of life elsewhere in our solar system. Such arguments have assumed that if bacteria can live in near-boiling water, in rock fissures deep underground, or within frigid Antarctic ice here on planet Earth, *then it is reasonable to expect life forms* in similar environments elsewhere in the universe.[29] Nonsense! The flaw in these arguments is that they ignore the problem of life's origin and earliest elaboration. Bacteria living in Earth's extreme environments have had three billion years of competition to end up where they're at; it seems highly unlikely that they *originated* in these extreme habitats.

FOSSIL EVIDENCE FOR A MORE COMPETITIVE WORLD

Paleontologist Geerat Vermeij has examined what he calls **Evolutionary Escalation** by studying the rich fossil record of mollusks. Unlike most shell collectors, Vermeij has been especially interested in those specimens that were broken, damaged, or repaired during their lifetimes. Blind since childhood, Vermeij studies shells with his sensitive fingers and became fascinated by the kinds of damage he encountered. By studying damaged shells in the fossil record, he examined how some mollusks have changed over geologic time. In his book on escalation, Vermeij proposed several hypotheses to test the notion that competition and predation have played an important role in evolutionary trends.[30] First, he hypothesized that competition and anti-predator capacities of individual lineages have increased over time. Second, he suggested that recently evolved individuals should be better adapted to a more hazardous environment than their earlier ancestors. Third, he predicted that, as time moves forward, hazards in the environment will have become more severe. Vermeij tested these conjectures with examples from many sources, but the damaged shells of fossil seashells were his primary data set. As the crushing pincers of some crabs have become larger and more powerful over millions of years, seashells—their prey—have developed thicker, stronger, and more elaborately ornamented shells. (The survival of many seashell lineages with thinner shells, or crabs whose claws have not become enlarged, does not negate the fact that a few lineages were part of an "arms-race.") Consistent with both Vermeij's "escalation" and Van Valen's "Red Queen," the fossil record supports the notion of increasing predation intensity over time.[31]

Competition really does appear to be a critical element in directing evolutionary trends. Our antibiotics to fight bacterial infections are a fine example. Penicillin, streptomycin, and similar compounds were first derived from Actinomycetes. These are bacterial-grade microbes that live in the soil, breaking down organic matter for their sustenance. The *reason* they produce powerful antibacterial substances is simple: bacteria are consumers of the very same resources. What better way to survive than to poison your competitors? These same "poisons" have become our **antibiotics**, saving millions of human lives.

The sudden expansion of mammals after the end-Cretaceous extinction has been explained as a *release from intense predation* after the dinosaurs had been eliminated. Once those fearsome beasts were no longer part of the landscape, mammals diversified explosively. Here, a change in the competitive landscape allowed an *underclass* of smaller furry animals to proliferate. A similar scenario occurred after South America became linked to North America via the new Panamanian land connection around three million years ago. Surrounded by oceans for more than 50 million years, "Island South America" had developed a very peculiar mammalian fauna. Sloths, anteaters, armadillos, neotropical monkeys, and a rich variety of smaller marsupial mammals (such as the opossum) are living representatives of South America's original fauna. Extinct giant ground sloths, large ungulate herbivores, and huge armadillo-like creatures were also part of South America's distinctive fauna. But with the "Panamanian bridge" suddenly available, land animals were able to move between the two continents. Two distinct faunas came into direct contact, giving rise to what has been called "the great American interchange."[32] Giant ground sloths, opossums, armadillos, and monkeys moved from the south into North America. These animals added to the rich fauna of the north, but the story on the other side of the *exchange* was different. From the north came a more lethal crowd. Wolves, bears, large and small cats, raccoons, deer, camellids, and various rats and mice made South America their new home. For many indigenous South American animals these new neighbors proved deadly. A distinctive fauna was no longer protected by its isolation, and many of its unique animals became extinct. Initially both North and South America had about twenty-six families of mammals. However, after the intercontinental exchange, 50 percent of South America's original mammalian genera were gone, while North America lost around 28 percent. (Ice age fluctuations account for the northern losses. But then, beginning around 12,000 years ago, humans with a new hunting technology eliminated many large mammals throughout the Americas.[33])

Among neotropical plants, there is no evidence for extinctions after South America became connected to North America. Clearly, among animals, competition plays a more critical role in fashioning the biota. This is especially apparent on isolated islands, where less competitive animal lineages still survive, whether marsupials and monotremes in Australia, or lemurs on Madagascar. Isolation provided protection. All of South America's endemic ungulates became extinct after they were challenged by a more aggressive northern fauna.

COMPETITIVE ARENAS: LARGE AND SMALL

What made the North American mammals more effective competitors than their southern neighbors? Why have island animals been weaker competitors than continental invaders? The answer appears to be: *more intensive competition over a larger land area for a long period of time.* Contiguous Africa, Eurasia, and nearby North America provided an immense "playing field" on which competitive scenarios could play themselves out. We've seen much the same thing in more recent times, as the introduction of Eurasian plant diseases wiped out the American chestnut and decimated American elms. Worst of all was the loss of Native American people after Columbus had opened the New World to European immigration. With the introduction of human diseases from Eurasia and Africa, indigenous American populations suffered huge declines. Native Americans had not been part of the "arms-races" between humans and their pathogens ongoing in Africa and Eurasia over the previous 15,000 years. Isolated from these nasty interactions, Native Americans lacked the immunities that people in the Old World had acquired over this time span. During the two hundred years following the discovery of the "New World," population decline among Native Americans may have been as high as 80 percent. Whether for North American trees or Native America peoples, the scenario was the same. The larger contiguous Old World land masses, with continual pathogenic interactions, provided a more harrowing environment than a smaller isolated land area. When the survivors of the larger war encounter those who have not undergone this same trial by fire, the former prevail.

The take-home message from these various tales is clear: competition, whether in the form of disease, predation, or crowding, has been a driving force of evolutionary advance. A majority of species formation appears to be driven by niche division and ecological differentiation (chapter 2). This too is a direct consequence of competitive interactions, where peripheral populations avoid competing with others of their own kind, by entering territories new to members of their own species. Similarly, the pioneering effect drives populations into ever-more challenging environments. Competition results in a world of increasing crowding and nastier neighbors. Nevertheless, some species can "facilitate" the survival of other species, whether weeds beginning plant succession or insects pollinating flowers. Together, these many factors give us a world that is both dynamic and has become *progressively more complex*. (This view is not an artifact of the "hindsight fallacy" where, looking backward, we seem to see directional trends.) Let's consider two grand triumphs in the history of increasing biological complexity: social insects and mammals.

A NEW LEVEL OF COMPLEXITY: THE SOCIAL INSECTS

As we discussed in chapter 1, those insects with a four-stage life cycle have amassed the highest species numbers within the animal kingdom. By dividing their lives into a start-up stage (the egg), an eating/growing stage (the larva), a transformative stage (the pupa), and a dispersing/repro-ducing stage (the adult), these animals have become the most numeri-cally abundant on the planet. But now, let's switch from the number of species to the number of individuals.

In chapter 6 we saw how between-species cooperation could lead to greater overall biodiversity. Whether fig trees and their wasps or soil fungi and plant roots, cooperative symbioses have helped many species live together successfully and multiply. But there is another form of coop-eration we didn't mention: *within-species cooperation.* No single wolf can hunt with the effectiveness of a pack of wolves; no single ant can do what a colony of ants is capable of. Social species are yet another advance in biological complexity. The most advanced of these are **eusocial** animals, living in multi-generational communities with a stratified division of labor. Here, individuals sacrifice their own interests—even their lives— for the good of the community. Leaving our own super-social species for the final chapters, let's examine the social insects.[34]

Eusocial insects have non-reproductive castes within their communities to perform specific tasks. The best studied example of this lifestyle is in honey bees (*Apis melifera*), where sterile worker bees maintain the hive, nurture the young, and gather food, but do not participate in reproduction themselves. The young worker bee begins by nursing the larvae, becoming an active forager only later in her life. Here is a clear division of labor within the life span of the individual worker—all in the service of the colony's welfare. The beehive and its queen make up a single entity, often called a **superorganism.** Other eusocial insects include termites, some wasps, some bees, and many ant species. Though eusocial insects comprise only 2 percent of described insect species, surveys suggest that they make up a majority of the world's actual insect *biomass!* In one Amazonian rain forest survey, 75 percent of total insect biomass was estimated to be made up by colonies of eusocial insects.

Found only in neotropical forests, leaf-cutter ants carry leaf fragments down from high trees into their underground nests, nourishing the fungi they feed upon. Large underground colonies can number as many as five to ten million ants. Their leaf-gathering activities are often clearly evident on the forest floor, where narrow trails, free of debris, lead to favorite trees. (They remove litter from their pathways!) Leaf-cutter fungus culture is estimated to have evolved from debris-gardening ants around ten million years ago.[35] These little ants, thanks to their numbers and their industry, play a significant role in the life of lowland neotropical rain forests.

Astounding: only two percent of insect species, but accounting for more than half of insect biomass! Stratified social hierarchies have paid off for these lineages. Just as Adam Smith declared: *the division of labor provides economic efficiencies.* Among some ants, aggressive "soldiers" are ready to sacrifice their own lives for the protection of the colony—a very effective way of maintaining the fitness of the group. But why is eusociality so rare? The hurdle appears to be that a large percentage of individuals must become genetically reconfigured to abandon their own reproductive potential for the "greater good" of their community—and this doesn't come easy.[36] Eusociality seems to be a bold contradiction to ordinary natural selection, where the reproductive success of each individual is the object of selection. Special preconditions in the behavior, genetics, and lifestyles of these animals were necessary before eusociality became possible.[37] Here, the fitness of the individual is sacrificed for the greater success of the community, in what has been called **group selection** or **multilevel selection**. The cooperative community is the unit-of-selection and not the individual!

The hive or the colony stands or falls on the effectiveness of all its castes working together for the common good. Despite their little brains and limited behaviors, natural selection has given rise to a "swarm logic," making these insect societies remarkably resilient. While rare, this strategy of increasing social complexity and stratification has proven successful over a wide range of habitats. Not a big jump in species numbers perhaps, but surely a large increase in numbers of individuals.

Termite mounds are often a conspicuous feature in tropical grasslands, where they play a role in soil dynamics and where termites make up a significant portion of the living biomass. Eusocial ants are numerous and important around the globe. From a broader perspective, the elaboration of insect societies is another fine example of how our living world has become increasingly elaborate over recent evolutionary time. But while insects are the most numerous of all land animals, they are not the most dominant.

ANOTHER PARAGON OF LIVING COMPLEXITY: THE MAMMALS

Warm and furry, mammals are a marvelous and unusual crowd. Numbering around 5,500 living species, these animals are found in even the most severe terrestrial ecosystems. Their unusual success is due to a variety of factors. Most notable is a warm covering of fur, especially important for little mammals that lose heat rapidly because of their larger relative body surface. Consistently warm body temperatures allow mammals to be active over a wide range of external temperatures, metabolize more rapidly, and respond immediately to challenges. Maintaining higher temperatures, however, demands a steady supply of nutritious food. This, in turn, required the differentiation of teeth able to cut, tear, and grind. In addition, all mammals share a uniquely complex inner ear. Better hearing may have been essential for living a nocturnal lifestyle during those times when quick carnivorous dinosaurs ruled the daylight hours. Mammals are further distinguished by mammary glands, providing high-quality food for their newborn young.

The earliest mammals diverged from mammal-like reptiles about 200 mya, and produced a variety of lineages over the ensuing 150 million years.[38] Like many other lineages, mammals show some increase in diversification as the flowering plants expanded, around 80 million years ago.[39] Today's most primitive mammals are the monotremes, still laying eggs, and with their arms and legs splayed out sideways in a reptile-like configuration. More advanced living mammals fall into two groups: the marsupials and the placentals. Marsupials give birth to small, poorly developed young that must crawl to the mother's teats, where they attach themselves and begin nursing. The reason the young are born so small is that they must escape the uterus *before the mother's body produces antibodies against them.* Since the little ones contain genes from their father, they are, in effect, foreign bodies within the mother, and the mother will produce antibodies that can kill them. Developing largely while nursing has constrained the variety of marsupials. Nevertheless, some marsupials resemble mice, a few look like dogs or little bears, and some become quite large, hopping around on their hind legs. Overall, the marsupial lineage has been limited in its morphological diversity—all because the little ones must escape the womb early in their development. Not so the placentals!

One of the most sophisticated inventions in the history of animal life has been the **placenta.** This intricate organ allows the blood vessels of the mother to transfer food and oxygen to the blood vessels of the fetus, while accepting carbon dioxide and waste products from the fetus. Thanks to the complex placenta, fetal growth transpires *without triggering antibodies* in the mother's own body. What this means is that a mother elephant can carry her calf for twenty-two months before giving birth. It means that the fetus can develop, fully protected, within its mother's body for however long it takes. Newborn whales and porpoises come to the surface for air and swim alongside their mothers immediately after they are born. Newly born antelopes can run quickly only an hour after birth. Once the dinosaurs departed, the complex placenta has allowed placental mammals to diversify into a huge variety of forms.[40] The smallest mammals, little shrews and the tiniest bats (weighing around two grams, or 1/16 oz.), and the blue whale (reaching up to one hundred tons) are all "Placentalia"! No other animal lineage comes close to being as diverse in size and form as do the placental mammals. All thanks to a placenta that effectively prevents the mother's defense system from destroying the fetus she is carrying.

Another advance in complexity is found in the ruminant mammals with a four-part stomach and the ability to regurgitate their food so that it can be chewed again for further digestion. Because of the many microbes living in their extended alimentary canal, these are among the few mammals that can derive energy from the digestion of cellulose. While all mammals have microbial symbionts helping them digest their food, ruminants are the most numerous large herbivores in the modern world.

Returning to the placenta, one of its most important benefits has been the elaboration of larger brains. Like so many other "experts," behavioral scientists disparaged animal intelligence over many decades. Any attempt to describe mammal behavior in human terms was derided as *anthropomorphism*. In contrast, Darwin was deliberately anthropomorphic in his descriptions of animal intelligence and their expressions. Fortunately, recent work, ranging from parrots to chimps, has brought us back to Darwin's perspective. Here's a recent report I think revealing: It seems that "Rico," a German family's pet border collie, had been given many toys over time, each toy with a specific name. When scientists heard that Rico could retrieve two hundred different toys by their two hundred names, they decided to investigate. Rico and the scientists were situated in one room, as Rico was instructed to fetch a specifically named toy from another room holding about ten different toys. Rico proved to be correct 95 percent of the time. However, the scientists varied their procedure, and they did this several times. In these instances, they told Rico to retrieve a toy whose name *he had never been taught*, and Rico returned with *a toy he did not know.*[41] This clever canine knew what he did not know! (Really smart, seems to me!) But why should mammals have been enlarging their brains over time?

THE NATURE OF INCREASING BIOLOGICAL COMPLEXITY

While the history of life has been one of increasing numbers, the occasional input of greater complexity has provided new platforms for further diversification. Though the fossil record is best preserved in marine sediments, it is on the land where both species numbers and biological complexity have reached their peak.[42] Sculpted by **natural selection** imposed by an unforgiving and dynamic environment, further propelled by incessant competition, we now find ourselves in a gloriously diverse biosphere.[43] But what might be the underlying mechanisms that foster increasing complexity?

In an essay titled "On the Evolution of Complexity," W. Brian Arthur proposed three mechanisms that drive increasing complexity, whether in biological, economic, or cultural systems.[44] He calls his first mechanism **Growth in Coevolutionary Diversity**. Here entities coexist in interacting populations, and these interactions can drive the system into greater numerical richness. Within-species competition may drive some individuals into peripheral isolation, where new species are likely to form. New species, in turn, create new niches for others and enlarge environmental complexity.

Arthur calls his second mechanism **Structural Deepening**. Here entities can break out of their earlier limitations by adding functions or subsystems that improve their competitive or adaptive abilities. One can imagine how the invention of flight opened myriad possibilities for insects. For humans, learning how to walk on two feet allowed them to do new and amazing things. Arthur argues that competition drives these mechanisms ever forward.

Finally, Arthur proposed **Capturing Software** as a third mechanism, driving rare but sudden escalations in complexity. Endosymbiosis, whether with mitochondria or chloroplasts, embodied the "capturing" of new abilities. Recently discovered, a unicellular red algal species (a Eukaryote) has acquired genes from both Bacteria and Archaea in order to survive as an extremophile.[45] In our own species, the "software of language" accelerated cultural innovation. Today, scientific methodology drives technological advance ever forward.[46]

Increasing complexity appears to be universal. In a short review of super conductivity, physicist Dirk K. Morr states that: "The emergence of complexity is often tied to novel forms of collective behavior driven by strong interactions."[47] Whether in condensed-matter physics or exploding stars, our universe has become richer in complexity over time. Within a living cell, complexity allows redundancy and dynamic networks to maintain equilibrium in the face of both internal and external challenges. Over the history of life, competition has made survival increasingly difficult. As a consequence, and only in recent times, one lineage has responded by becoming *very, very smart!* Thanks to our agricultural symbionts, we humans became increasingly eusocial, relaxed the selective pressures that constrained us, and made possible increasing social complexity.[48]

In the final chapters, we will examine two of the world's most extraordinary examples of advancing complexity. The first was biological: the ascendance of human intelligence. The second has been a direct consequence of this same advance, but something entirely new and dramatic: the elaboration of human cultures. Moving from small pagan chiefdoms to strong unifying theistic religions allowed large societies to maintain close internal cooperation. Such societies drove the purposeful advance of human technologies.[49]

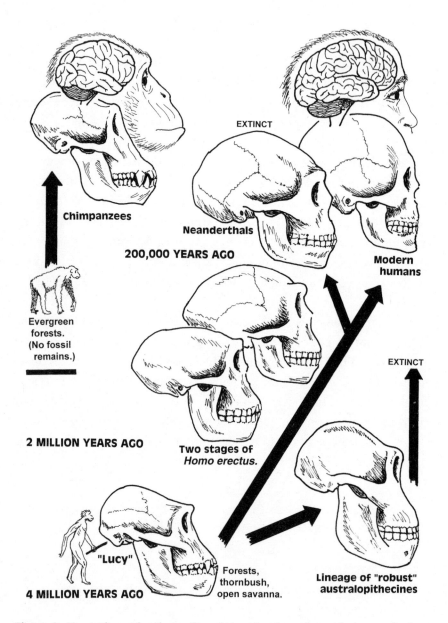

Figure 2. Hominid cranial evolution. "Lucy" (*Australopithecus afarensis*) had a cranial volume of about 400 cc, similar to modern chimpanzees. Average cranial volume expanded from about 800 cc in early *Homo erectus* to as much as 1,500 cc in Neanderthals, and 1,300 to 1,400 cc in modern humans.

Chapter 9

BIOLOGICAL COMPLEXITY TRIUMPHANT: THE HUMAN MIND

The Earth is a garden in space and we are some of its blooming life.
How odd to be a sack of chemicals that can contemplate itself, and
how much fun.

—Diane Ackerman[1]

For thousands of years, we humans ascribed Earth's rich biodiversity to a Creator who had both designed and fashioned the world around us. In the Judeo-Christian tradition, this grand creation had taken place about six thousand years ago. Later, medieval Christian philosophers suggested that we might understand God better if we were to investigate his handiwork directly through the study of nature. After all, if God had given humans commandments to properly live our lives, perhaps God had provided nature with laws as well. Indeed, in 1687 Isaac Newton proposed a simple mathematical framework—gravitation—to explain why an apple falls to the ground and how the planets circle our Sun. Natural philosophy made clear that God's universe was both lawful and comprehensible!

Biology and geology, rather more intricate than planetary motion, took longer to reveal their inner workings. It was not until the early 1800s that Earth's long history was elucidated. The importance of cellular organization in biology became apparent at around the same time, while the nature of disease was not understood until the late 1800s. Darwin had proposed a natural mechanism for the origin of biological design in 1859, but his ideas were not successfully integrated with genetics until the 1930s. In 1953, the double helix of DNA was identified as the mechanism for replicating hereditary information and carrying this information across generations. Now, with an understanding that fruit flies, mice, and people all use similar genes to build themselves, the glory of morphological development is yielding its secrets. Science has given us an **evolutionary epic** spanning almost four billion years of biological history—an odyssey of increasing complexity over time.

Ecosystems themselves have become richer in species and more complex over time. Terrestrial vegetation and the trophic levels it supports are far richer today than in ages past. The great coal deposits of the Carboniferous period were the consequence of having far fewer decomposers than today's tropical forests. Over the last hundred million years, flowering plants have further expanded terrestrial species richness, both with their own numbers and with those who feast upon them.

One of the unusual features of flowering plants is that they appear to spend less energy on building defensive chemistry than do other seed plants. The result is that they can sustain many more herbivores. Peter Ward claims that the world's biomass may have peaked over 300 million years ago (mya).[2] Very unlikely, I believe, considering what's been happening on land surfaces! By developing a vegetation far richer in both species numbers, structural variety, and nutritional resources, flowering plants created new environments in which insects, mammals, and birds have also proliferated.[3] Best of all, and from our own perspective, these increasingly more nutritious environments allowed one particular animal lineage, the primates, to build bigger brains. All told, this is a story of progressive advance. However, for many scholars, the notions of progress and increasing complexity over time have been seen as highly suspect, culturally biased social constructs.

THE CONCEPT OF PROGRESSIVE CHANGE

> *Natural selection therefore works like a ratchet, which turns the operation of random variation into a trajectory.*
>
> —Nick Lane[4]

Paleontology—the study of fossils and their history—paints a picture of escalating numbers and increasing complexity over time. This differs little from our views regarding human cultural history. First we were hunter-gatherers, then settled agriculturalists, and soon building grand civilizations. Though many larger city states may have risen only to collapse, human knowledge and technology have continued to grow and expand over time. With dramatic innovations changing people's lives during the nineteenth century, the *notion of progress* became a central tenet of Western thinking. Since the science of paleontology expanded as the Industrial Revolution unfolded, progressive change was understood to have begun at the very dawn of life.

Nevertheless, and for over much of the last century, discussions of *progress* have been unwelcome in the biological sciences. Stephen Jay Gould was especially disdainful of progress in any of its manifestations.[5] He argued that natural selection is totally devoid of any progressive drive and, in this view, he was perfectly correct. But here again we stumble into nature's many-tiered complexity. Just because individual reproductive success is not, in and of itself, progressive does not mean that progressive trends cannot arise along the way. Accidental gene duplications allow for increasing developmental complexity; predation pressure drives counter-adaptation. From a broader perspective, **progress** appears to be an *emergent consequence of evolutionary dynamics*: the inevitable consequence of many interactions over time and space.

Academic hand-wringing over whether there really is such a thing as *progress* in the history of life continues to be fashionable. Writing about verifiable trends in the increase of complexity over time, Daniel McShea notes, "Given the historical background and the power of culture to penetrate perception, it is reasonable to wonder whether this impression of large-scale directionality is anything more than a mass illusion."[6] He himself presents evidence for increasing complexity within the vertebral column over time—from fish to human—but worries that trends of simplification can "offset" such increases. Humbug! Imagine a world overrun by simplified parasites. Would this erase the fact of increased complexity in other lineages over time? I don't think so.

Looking backward into time, and impressed by the rise of successful lineages, it is easy to see "directionality" in much of life's history. Looking forward, however, provides no way of discerning future directions; there is no way of predicting success or failure. Worse yet, extinction is the inevitable endpoint for most lineages. Centuries ago, natural philosophers had envisioned God's purpose at work in the slow unfolding of the natural order, with our own species (*Homo sapiens*) as the culminating glory in this odyssey. Such anthropocentric self-adulation has been banned from the halls of modern science. While abandoning a deity-directed progression of life, scientists also came to avoid the concept of evolutionary progress. However, if we accept the Big Bang scenario, the universe itself has progressed mightily. Starting with hydrogen, helium, and a bit of lithium, we are told that aging and exploding stars helped forge all the heavier elements now gracing the cosmos. Surely beginning with three elements and ending up with more than ninety represents a profound increase in complexity. A few of these heavier elements sit inside our most essential enzymes, absolutely critical in maintaining complex life forms. Clearly, the history of both our universe and ourselves has entailed progressive advance. Since Darwin's time, the metaphor of progressive evolutionary change has proven fundamental to our understanding of the living world.[7]

The idea of progress is supported by studies of fossil marine lineages; in many genera, extinction rates have tended to diminish with time. Many genera—as they move through time—produce more species, and these species tend to cover larger geographical areas or invade a wider variety of ecological niches. Such trends should diminish the probability of generic extinction over time. In addition, genera having special qualities decreasing their extinction rates should also increase over time.[8] This is natural selection operating beyond the level of individual species.

Many scholars conflate the word *progress* with the notion of *betterment*. Progress, however, has wider meanings. When your doctor informs you that your disease is progressing, betterment is not what comes to mind! Though progressive change may be the icing on evolution's many-layered cake, there are exceptions. Speciation and lineage-splitting may not involve substantive increase in complexity. Such trends give us many more species, but no new organs, no new abilities. This is diversification in a "lateral," not an "upward," direction.[9] In addition, selection can result in evolutionary stasis, where a lineage persists over long time periods without evident improvement. As boys exploring Long Island's Great South Bay, my friends and I were fascinated by horseshoe crabs (*Limulus polyphemus*). Some were up to two feet long. And all had a long spike for a tail. Resembling fossil trilobites, their front end is a single rounded U-shaped carapace with two wicked-looking eyes on either side. We would usually encounter their translucent exoskeletons, shed when they molted. Once in a while, we engaged with the living beast along the shore. But here's the point: a recent fossil discovery suggests that horseshoe crabs—in much their present form—have been around for 400 million years. That is a very long time, and not very progressive. More amazing is the fact that we can put a human gene into a bacterium, in order to have that bacterium synthesize human insulin—an astonishing example of how deeply conservative some biochemical functions really are!

To make matters more confusing, there are animals that combine highly "advanced" traits together with "primitive" characters, all within the same body. The rear end of a platypus has a cloaca, rather like reptiles, from which females discharge a leathery egg—just like reptiles. But the front end of the duck-billed platypus is an extraordinarily sensitive device, highly attuned to seeking prey in muddy water. In addition, and almost unique among mammals, the male platypus has poison glands! Here we find ancient and advanced traits in the same mammalian body. All told, evolutionary history is full of both conservative stasis and dynamic advance. But there are also parasitic lineages that have become much less complex. Using their hosts as resources, parasites can jettison a lot of excess baggage. Such exceptions do not negate a general pattern of progress in the history of life.

Philosopher of science Michael Ruse argues that a strongly progressive philosophy of human history, developed in the eighteenth and nineteenth centuries, has both inspired and debased evolutionary biology from its earliest beginnings. "Complexity is a poor guide [for assessing progress] since ancient forms (trilobites) were more complex even pos-

will follow Harry Jerison, who compared brain/body mass ratios.[14] Ratios are a better metric than simple brain size because ratios avoid the "big animals got bigger brains because they're bigger animals" conundrum. (Elephant brains are about three times as large as ours.) Comparing vertebrate animal brain/body mass ratios across lineages, Jerison found three differing data sets. First, birds and mammals sit well above the averages of brain/body mass ratios found in fish, amphibians, and reptiles. Secondly, primates stand a bit above most other mammals, and, third, we humans occupy a further step beyond the primate norm.

The historical escalation of brain volumes is one of the clearest examples of increasing organic complexity in the fossil record. But before we discuss the fossil record, let's take a minute to make clear what we mean by **intelligence**. Information processing is what brains really do. A pragmatic definition states that intelligence is the "ability of an organism to sense its surroundings and make appropriate responses." An appropriate response will help the organism find food, avoid predation, recognize a potential mate, and deal with other challenges. Complex animals with eyes, ears, and other sensory organs linked together in a neural network do the job of monitoring their environments. Animal brains take this sensory input from many organs to build representations of the outside world. Dragonflies, darting back and forth in their pursuit of mosquitoes, are a superb example of eye-brain-flight coordination, and they do this with a brain the size of the head of a pin!

In discussing higher intelligence, however, we are focused almost entirely on animals with backbones. Early in their history, fish developed an insulating fatty (myelin) sheath around their nerve fibers, allowing electrical signals to zip quickly along those fibers. Insulated nerve fibers allowed these early vertebrates to react more quickly to the challenges they faced. However, just as there is more biodiversity on land, brain power expanded more grandly on land. Clever dolphins and smart whales are mammals that first evolved on land, returning only later to the sea. The smartest invertebrates, eight-legged octopi, are marine creatures—but they are few in kind and number. Those few fish species that do have large brains generate electrical fields to seek their prey in murky water, using brain power to analyze the data they perceive. Despite these exceptions, larger brains are mostly found in birds and mammals. Why should this be?

Simply stated, building and maintaining a more complex structure requires more energy. We humans use about 20 percent of our resting energy just to keep our brains going. When we are deprived of oxygen, our brain is the first organ to suffer damage. That's the maintenance issue, however there is also the problem of *building* a bigger brain. Human infants utilize about half their food intake to expand their growing brain in the first year of life. In addition, maintaining a larger brain means burning more energy faster. Mammals require five to ten times as much food to sustain themselves as a reptile of the same weight.[15] Your pet lizard can survive a month without food; your pet gerbil will not. As we've noted, sharply differentiated teeth allow mammals to cut, tear, and chew their food, allowing more rapid digestion. Many toothless birds fill their crops with pebbles to grind up their chow. Both mammals and birds have longer intestines than reptiles, digesting their food more quickly, and both have a four-chambered heart to distribute oxygen more efficiently throughout their bodies. In addition, mammals and birds are warm-blooded, keeping their bodies at higher temperatures, where metabolism is quicker and brains work more efficiently. Just as significantly, both birds and mammals must support the early development of their young. Surely building a bigger brain is a bigger expense. And yet the fossil record indicates a steady increase in relative brain size among many mammalian lineages over the last fifty million years.

Ethiopia's ape-like fossil "Lucy" (*Australopithecus afarensis*) lived around 3.3 million years ago and had a brain volume of about 450 cc, similar to that of a chimpanzee. (A cubic centimeter or cc equals 0.061 cubic inch.) Over the following three million years, our lineage expanded that brain volume to the one we're carrying around today: averaging around 1,400 cc. Here's a *three-fold increase in only three million years!* Surely one of the most dramatic examples of escalation in the fossil record, this scenario has an additional twist. Our fossil cousins, the Neanderthals (*Homo neanderthalensis*), were isolated from other humans in seasonally frigid Europe and western Asia for half a million years. During that time, they also evolved big brains! Their skulls are longer across the top, while ours are more rounded. Recent analyses of Neanderthal newborn skulls indicate that their brain growth was similar to our own.[16]

In addition to being brainy, contemporary humans exhibit a grand variety across the globe, while having considerable genetic diversity within each and every hamlet. Our species is hugely polymorphic! Unfortunately, paleoanthropological practice assumed that our ancestors were much less varied, giving new and different names to many different fossil remains. This practice has populated our past with a "shrubbery of species." From my readings, it seems that *Homo erectus* may have been as variable as we are, and can serve as most everybody's ancestor between one and two million years ago. A group of five crania recently excavated in Dmanisi, Georgia (including a fully preserved large male skull), support this view.[17] However, and with time, something new and different did arise in Africa, perhaps 200,000 years ago. This lineage had a more slender frame, a higher forehead, a less protruding chin, smaller teeth, and was the first hominid to *throw spears with stone spear points!* We've given this distinctive form our own Latin name: *Homo sapiens.*

Spreading out of Africa and around the world, this is the lineage that provided most of the genetic heritage we humans carry around today.[18] However, even as Neanderthals were on their own trajectory in Europe, something very special was happening on the island of Flores in Indonesia. There, a population of early hominids adapted to their island home by becoming much smaller. Nicknamed "Hobbits" by their discoverers, these short, small-brained people lived as recently as 18,000 years ago and appear to have become distinctive (called *Homo floresiensis*).[19] These fascinating "little people" are a reminder that the evolution of humans has included a lot more *diversity* than we imagined earlier.

Returning to the two main strands of modern hominid evolution, both *H. neanderthalensis* and *H. sapiens* enlarged their brains significantly over the last million years. Neanderthals expanded their brain power in ice age Europe and western Asia, while our ancestors, *Homo sapiens*, expanded their brains in Africa and spread out from there. Despite the huge costs in food procurement and loss of mothers in birthing, both modern humans and Neanderthals increased their brain volumes independently! Here is one of the most extraordinary examples of escalating complexity in the fossil record, and it was done in parallel by two closely related lineages in very different environments! How did this unusual trajectory begin?

WHY PRIMATES GOT SO SMART

Early mammals had a brain/body ratio equivalent to that of today's opossums and hedgehogs, about three times the weight of the reptilian average. For most mammals, this level of brain enlargement (encephalization) was maintained well after the dinosaurs went extinct (65 mya). Fifty million years ago, the dog-sized ancestor of horses (*Hyracotherium*) had a brain/body mass ratio similar to that of modern opossums. Becoming larger over time, horses reached the average modern mammalian brain/body mass ratio by twenty million years ago. Luckily for us, there was one mammalian lineage that did not fit these general patterns.

From their beginning, primates tended to have larger brains. While most mammals allocate about 5 percent of their metabolism to maintaining brain function, primates utilize 9 percent or more. What made the monkey lineage smarter? The initial step may have been shifting from a nocturnal lifestyle on the ground to the daytime pursuit of nourishment in treetops. With the expansion of the flowering plants, trees bearing flowers and fruits became an important resource for a variety of animals. Pursuing insects and fruit in the high canopy, early arboreal primates required better three-dimensional vision and more physical agility. Over millions of years, muzzles became shorter and eyes came closer together, allowing visual fields to overlap. A larger brain now processed two slightly incongruent images for accurate depth-perception, important when jumping from branch to branch. Binocular stereoscopic vision had additional benefits, such as finding tasty but camouflaged bugs to eat or watching for predators in the dead of night.[20]

Because their lives were now spent among the tips of branches, where blossoms and fruits reside, primates evolved flexible arms and fingers to catch insects and reach for yummy fruits. Longer rear legs with grasping toes helped hold on tight. Separate digits, with a flexible thumb and big toe, were important in grasping slender twigs. Claws became transformed into more useful nails, while the grasping surfaces of hands and feet developed a ridged skin, helping hold onto smooth branches, and giving us our fingerprints. At the same time, arms and shoulders were developing new kinds of flexibility. Hand-to-mouth feeding—after careful examination—is characteristic of all primates. Surely, none of this monkey-business would have been possible without the flowering plants. Neither tree ferns, cycads, nor conifers provide such a variety of nectar-filled flowers, juicy fruits, or nutritious seeds as do the flowering plants. These features attracted many insects into the treetops, quickly followed by the earliest insect-eating primates.

With forward-pointing eyes, primates lost much of their lateral vision. Thus, it became important to travel in small groups, where extra pairs of eyes could survey all directions, watching for eagles, snakes, and arboreal cats. These social interactions promoted increased brain size. The better you understand the intentions of your associates, the better you can interact with them and make the troop an effective social unit. In general, the larger the number of interacting females in the troop, the better developed is that species' neocortex.[21]

> *At all stages of pregnancy, a primate fetus consistently has about twice as much brain tissue as a similar sized fetus of any other mammal. In other words, brain development is specifically privileged in primates.*
>
> —Robert Martin[22]

As primates continued evolving, there were two instances where brain size expanded even further. The first increase took place around twenty million years ago, as seen in the fossil called *Proconsul*. These animals were among the earliest of the apes. Not only did *Proconsul* have a slightly larger brain, it lacked a tail, as do all the living apes (the hominoids: gibbons, orangs, gorillas, and chimps). Spending more time in an upright position, both legs and back became stronger. More significantly, apes became the only mammals to be able to rotate their forearms rather like a windmill, allowing them to *swing through the canopy*. This is another instance where flowering plants made a singular contribution to our ancestry. Few non-flowering trees have broadly spreading branches in their upper crowns. Flowering trees do, giving rise to tropical forests with a canopy of many spreading branches, where a larger animal can swing from tree to tree. Called brachiating, our *swinging ancestors* developed broad shoulders, long rotatable arms, flexible elbows, and strong hands. Flowering trees not only fostered the origin of primates, they are responsible for our remarkably flexible arms!

WALKING ON TWO LEGS

After the origin of apes, the next critical event for our lineage was spending more time on the ground. Chimps and gorillas spend a lot of time on the ground, and both have developed a four-limbed, knuckle-walking mode of travel when on the ground. Human ancestors instead became more upright and two-legged, probably sometime around six mya.[23] By three mya we have better fossil evidence, as represented by Lucy and her kin.[24] Though Lucy's brain was the size of a chimpanzee's, her skeleton makes clear that *she walked upright on two feet*. Her arms were still long and ape-like, indicating that she spent nights sleeping in trees. (You need to be careful where you sleep in the African savanna—what with lions, leopards, and hyenas prowling around.)

Walking on two feet—**bipedality**—was a crucial advance for the hominid lineage. Our feet, in fact, are the most distinctive element of our skeletons. No other primate has feet with little toes up front, a big toe alongside the others, a tough padded heel, and an arch that translates impact into forward propulsion.[25] Standing upright gave us a better view across the grass savanna, reduced our exposure to the noonday Sun, and freed our forearms from having to support us. The fact that Lucy had modern feet while still carrying around an ape-like brain makes clear that our upright gait came well before we built our larger brains.[26] A further advance, developed in *Homo erectus*, was a more flexible neck, allowing us to survey the landscape quickly, and longer legs to expand our range. Much earlier, our ancestors had abandoned the muscularity of chimps and gorillas to become lighter, covering ground ever more nimbly.[27]

Bipedality proved transformative, freeing our forelimbs to carry things around, throw rocks, and fashion tools. Pressing our opposable thumb against a fingertip gives us a strong "precision grip," useful for everything from basket-weaving to using stone flakes as if they were razor blades. Opposing our thumb against all our fingers gives us a "power grip" for wielding clubs and hammers. Just as significant, our hands were now free to signal to our companions. By using arms and hands to make gestures, then coupling those gestures with vocalization, humans gradually developed their premier talent: language!

> *Language is a trick that allows the mind to question itself; a magic mirror that reveals to the mind what the mind thinks; a handle that turns the mind into a tool.*
>
> —Kevin Kelly[28]

Our minds are marvelous devices; we use our brains so effortlessly that we fail to appreciate what's really going on inside. Our vision appears to be simple and straightforward, but it is amazingly intricate. Visual information-processing protocols begin within the eye itself, where the signals from a hundred million rods and cones must be collated and processed before being transmitted by an optic nerve having "only" a million nerve fibers. The image projected on our retina is upside-down; our mind makes things right-side-up. Years of experiments with domestic cats and small monkeys have shown how different batteries of neurons and different areas of the brain contribute to the creation of what appears as a simple moving image. Thanks to clever brain processing, *tilting our heads does not tilt the horizon!* We notice unusual movements in the environment immediately, even when they are at the edge of our peripheral field. The macula, a small central area where we see sharply, scans the moving element repeatedly as the brain analyzes the object. Failure to recognize a predator could be fatal, but over-reacting to gusts of wind wastes precious energy. Decisions need to be made quickly and economically.[29] Distinguishing the intermittent cadences of a gusting wind from the interrupted rhythm of a quietly stalking predator may have created in our mind those faculties we now use when making music or exploring the structure of mathematics. The "human mind" is the emergent property of a large and complex brain.

HELPLESS BABIES GAVE US LARGER BRAINS!

A critical step in escalating hominid intelligence came only after the bipedal stone-throwing troop could defend itself on the ground and through the long tropical night. Defense during the night and on the ground may have involved constructing encampments enclosed by a barricade of thorny branches. It was not until this level of *group defense* had been achieved that females could give birth to helpless infants; infants unable to cling to their mothers' bodies. Once on-the-ground-security was achieved, mothers were protected through the long and dangerous night. Though we have no fossil evidence, fiber-woven slings—holding baby close to mother—was an essential early innovation. Though the birth of helpless babies placed our ancestors at greater risk, these infants were born with their cranial sutures unfused. Because these sutures are open, baby humans double their brain size in the first year of life! This little detail, coupled with a much longer childhood, built our larger brains.[30]

The human brain is hugely expensive; though only 2 percent of our body weight, 20 percent of our resting energy is devoted to maintaining the brain. No other organ burns so much energy on a continuing basis. (Though sleep gives us some respite, nerve cells operate continuously; they cannot be turned off.) While expanding human intelligence involved costly increases in neurologic hardware and interconnectivity, there was an even greater cost. The human birth canal has lagged behind brain expansion. Tragically, women suffer the most difficult and dangerous birthing among vertebrates. Childbirth has been a dreaded killer of women throughout recorded history. Even in today's world, as many as 10 percent of Afghanistan's women are likely to die during or shortly after birthing. What forces of selection could possibly have countered so costly an expansion of the human brain? Whether in human economies or in natural selection, benefits must equal costs! What manner of selective pressures might have driven the costly expansion of our brain volume?

BUILDING THE LARGER HUMAN BRAIN

Not only are our babies helpless; born to healthy mothers they can be called "obese." Human females themselves, unlike any other primate, averages 20 percent of their body weight in fat, helping the fetus develop and nurse the infant during food shortages. In contrast, men's bodies average around 7 percent fat. In addition, and unlike other primates, the female pelvis differs considerably from that of the male, allowing for a larger-headed baby to pass through the birth canal. *Homo sapiens* is a **sexually dimorphic** species, with plump females having the essential responsibility of birthing and nurturing our young.[31] Males, on the other hand, must secure and defend territory with sufficient resources to sustain the group. In fact, dangerous birthing and helpless babies may be the reason why women experience menopause. Released from the dangers of further birthing, older mothers can contribute significantly to the care of their grandchildren. Basically, motherhood is the second most important job in the world; acquiring nutritious food is the first.

A troubling conundrum in biology has been the question of how Mother Nature actually goes about building something as complex as the human brain. This is a question very similar to asking how complex animals develop from a single fertilized egg, and the answers are much the same. Not too many genes, coupled with carefully cadenced developmental protocols, can create very complex, yet fully integrated, structures. The fact is that there simply aren't enough genes in the human genome to specify a brain having billions of neurons, all nicely interconnected by trillions of synapses. Genes do not carry sufficient information to specify the complex brain. Clearly, human brains *build themselves!* Genes may provide the necessary scaffolding, but developmental protocols and environmental cues build the basic structures. More than that, "synaptic plasticity" allows changes to be made in neural connections. Parts of the brain that are used expand and proliferate, while unused parts of the neural network wither in a process called "cognitive sculpting." Scientists came to these conclusions after doing some rather nasty experiments. By covering the eyes of kittens for a period of time during their early development, researchers discovered that these kittens would never learn to see. Though appearing perfectly normal, these kittens had been blinded! Visual input is necessary at a critical time in the kitten's development for the visual areas of the brain to form. *The visual system builds itself to see!*

Human infants deprived of social contact during their early years cannot learn to speak. Clearly, the brain uses external inputs to configure itself at critical points during development. Playfulness is characteristic of all smart animals. Play helps refine behaviors, practice social interactions, learn from accidents—all in preparation for the challenges of adulthood. With input from our environment and a kind of "natural selection of neurologic development," our brains *build themselves!*

But why might human intelligence have advanced in the face of huge energy costs and an appalling loss of mothers in birthing? The literature of anthropology has many suggestions and hypotheses explaining the escalation of human intelligence. Increased and more complex social interactions must have been important. Parents and their young cannot survive without a supportive social network. Born as helpless infants and with many years to reach adulthood, we require an altruistic social setting in which to mature. We are the only primates with white sclera around the iris of our eyes. In this way, we can see who is eyeing whom— clear evidence for our deep sociality. Fossil remains confirm that early humans had strong social bonds; our ancient ancestors cared for their afflicted brethren, both the toothless old and the compromised young.[32]

Unlike other primates, *human mothers share the care of their infants!* This explains why hunter-gatherer mothers averaged a birth every three years, in contrast to chimps and gorillas, where the birth interval is longer. Sarah Hrdy claims that a more socially interactive childhood made it imperative for children to understand the mental states and intentions of those around them.[33] In the words of David Barash, "The more conscious our ancestors were . . . the more able they were to modify—to their own benefit—other's impression of them. Those who possess an accurate theory of mind can model the intentions of others, and profit thereby."[34] Complex social maneuvering, coupled with cooperative child-rearing and shared intentionality, were critical elements in building successful human groups and an expanding consciousness.[35]

Tool use must have been another factor in driving the expansion of the hominid brain, especially the crafting of better hunting, cutting, and digging implements. Slings of woven fiber allowed mothers to carry their babies long distances. Surprisingly, early stone tools remained much the same over more than a million years, even as our brains were enlarging. Perhaps the gradual development of our most distinctive talent—**language**—was the driving force in the evolution of our deepening intelligence. Considering that tongue, mouth, and vocal tract had to work in exquisite synchrony with our mind, human speech undoubtedly developed slowly over thousands of generations.

A much less convincing argument for the cause of our escalating intelligence is that of climatic cycles during the ice ages.[36] If responding to climate change made humans smarter, why didn't other social hunters, like lions and wolves, also get smarter? More telling, how might climatic oscillations be responsible when two hominid lineages, one living in the cold of the north and the other in tropics and subtropics, expanded their brain volumes independently? None of the selective forces we just mentioned seem strong enough to counter the costs in energy, birthing, or early child-rearing that a large brain requires. Though rarely discussed in the literature of anthropology, there was another powerful "selective force" in the costly escalation of human intelligence!

TRIBALISM: HOW TERRITORIAL CONFLICT
MADE US REALLY SMART

For good and for ill, Homo sapiens *is inescapably a tribal animal.*
—David Berreby[37]

Though receiving little press, a singularly powerful factor drove the expansion of human intelligence. Excepting only pathogens and parasites, *competition with other humans* was the most dangerous aspect of our environment.[38] Not individual humans, but other packs of humans! As hunter-gatherers requiring a high-quality diet, we survived only in small groups. Local biomes, whether in African savannas or northern forests, can support only a few dozen mature humans and their young. Such environmental constraints kept our ancestors living in small bands over a very long time. Surviving on unreliable resources through each year and across the millennia was an incessant challenge.

Small independent human groups required safe shelter, drinkable water, and habitats providing sufficient nutrition. During a long drought, or simply with the expansion of our numbers, competition for limited resources became inevitable. For a local band of hominids, continued access to good habitat was the only way to avoid starvation. Competition between local tribes would determine who had access to good territory and who would be driven elsewhere—or expire.

Discussing the importance of conflict in human affairs has not been popular in the literature of anthropology. Introductory texts have surprisingly little to say about conflict or warfare; it's not part of the liberal imagination.[39] In strong contrast, a few fossil localities bear witness to a harrowing past. Fossil human bones mixed with those of tasty herbivores—all with similar butchering marks—make clear that, on occasion, our ancestors ate each other![40] Forget your modern sensibilities; survival was paramount. Moreover, **inter-clan conflict** is not unique to humans. It has been witnessed among chimpanzees, and is common in those other super-successful animals operating in small colonies: the ants. As their most incisive student, E. O. Wilson, pointed out: "The greatest enemies of ants are other ants."[41] And so it was for us.

In the context of selective forces driving the evolution of an ever-more powerful brain, the most significant factor was that humans were being stalked by an increasingly clever predator, operating in small packs. As Richard Alexander argued, "the only way to account for the striking departure of humans from their predecessors...is to assume that humans uniquely became their own hostile forces of nature."[42] Under such circumstances, one can see why language might play an especially critical role. Clear oral signals, alerting your comrades to approaching attackers, could spell the difference between life and death. A long history of inter-group conflict is consistent with an unusual aspect of both chimpanzee and human societies: we are *patrilocal.* Unlike gorillas and most other primates, human males usually remain with the natal group, where they become part of a defensive coalition. (Brides get exchanged; warriors do not.) Also, our clever brains work best when we operate in small groups of shared purpose. Whether hunting, gathering, or plotting, small groups of humans display an effective "collective intelligence." Indeed, defensive male coalitions with effective leadership gave rise to patriarchal societies around the world. Just as significant, inter-tribal conflict resulted in our becoming more altruistic!

> *A tribe including many members who, from possessing in a high degree the spirit of patriotism, fidelity, obedience, courage and sympathy, were always ready to aid one another, and to sacrifice themselves for the common good would be victorious over other tribes: and this would be natural selection.*
>
> —Charles Darwin[43]

> *Altruism would have facilitated the coordination of raiding and ambushing on a scale known in few other animals, while parochialism fuelled the antipathy towards outsiders. Additionally, with the development of projectile weapons, humans became adept at killing from a distance, which would have reduced the costs of aggression.*
>
> —Samuel Bowles[44]

Keen devotion to your comrades coupled with intense animosity toward your enemies is the only way of winning and securing territory. Getting yourself killed defending your clan may be necessary to the survival of your clan. (We draftees got this message in the US Army's Basic Training!) Behavior benefiting the group results in "positive group selection," despite the costs to individual fitness. It sounds ironic but it makes sense: greater conflict with outside groups promoted greater sacrifice within the group.[45] Females standing on the sidelines became part of the victor's spoils, further adding to group fitness. Surely quick and precise oral communication, shouted to your comrades in the midst of conflict, would provide a powerful selective advantage. More significantly, inter-clan conflict can account for the independent expansion of brain size in both Neanderthals and modern humans.[46] Though the ice age environments of *H. sapiens* and *H. neanderthalensis* were very different, by operating in small bands their behavior had to be the same.

Contrary to current fashion, I see our species as being *hard-wired for tribalism*, in much the same way as little humans are able to acquire language. We also carry "innate instructions" for love, friendship, and self-sacrifice within our community. There is absolutely no contradiction here; these are complimentary aspects of our innate **tribalism**! Over many millions of years, human survival required access to territory providing adequate food and water. Such territory was often limited, resulting in intense competition, and driving some groups into ever more challenging environments. Not only did inter-tribal conflict disperse us across continents, it made us the smartest and meanest critters on the planet!

Armed with a truly sophisticated computational device between our ears, and a highly versatile body, humans began elaborating their cultural skills. Language had given us rapid and sophisticated communication. Tool use became a major survival skill and was easily taught to others. The regular use of fire is documented as early as a million years ago in southern Africa.[47] By firing the landscape, early humans expanded grasslands, together with their protein-packed grazers. Sleeping around a campfire made the long African night less dangerous. Cooking over a fire was another significant advance, making food easier to chew and to digest.[48] Perhaps that's why both our teeth and jaw muscles became smaller over time, allowing our brain to expand more easily. By setting fires and hunting big game, humans began their transformation of the biosphere.[49]

Survival requires that cooperative systems be efficient. This meant that human males and females would come to play distinctive but complementary roles.[50] Birthing helpless babies and nurturing children over many years allowed human females to gather foods and fiber as well. Males hunted precious protein on long forays, garnering a resource essential for nourishing brainy little humans. Two very different sexes, both in form and behavior, gave our species a *division of labor* foundational to our success.[51] However, individual parents and their children cannot survive alone; they require a larger band to secure territory, share food, and provide continuing care in the event of a parent's death. Because of our need for energy-rich sustenance, humans survived only in small bands, closely bound together by altruistic cooperation. Born completely helpless and requiring many years of social nurturing to become effective adults, we became a deeply moral species. Regrettably, this morality was largely confined to members of our own tribe. Arguing for a wider *metamorality*, philosopher Joshua Greene states, "Once again, morality evolved (biologically) to promote cooperation *within* groups for the sake of *competition between groups*."[52] Incessant tribal warfare was the crucible from which human dominion has emerged; it continues under the banners of nationalism, religious factionalism, and ethnic conflict—just watch the news!

A NEW EVOLUTIONARY DYNAMIC: CULTURAL ADVANCE

A singular advance within our planet's long fossil record, humans tripled their brain volume over the last three million years. This splendid progression paved the way for new advances in terrestrial complexity. But it wasn't just a big brain that helped make our species the *master generalist* we have become. We are extraordinary in our physical abilities as well. We can walk, run, jump, and swim. Free and flexible arms allow us to carry things for long distances, beat with sticks, and throw rocks. No other animal on Earth can throw things the way we do, whether spears to take down game in times long past or a baseball at over 90 mph. Strong versatile hands with four slender fingers and an opposable thumb allow us to weave, whittle, sew, and fashion tools. Best of all, hands and flexible forearms allowed us to *gesture to our comrades*. This activity, coupled with a greater variety of vocal utterances, gave us **speech!** Verbal communication allowed us to warn each other over short distances, share thoughts within social networks, and formulate new tactics.

Operating in small bands, humans quickly invented new contrivances that would determine who prevailed and who fell by the wayside. Arriving in Eastern Europe around 40,000 years ago, *Homo sapiens* came armed with more versatile technologies. Fishing hooks and sewing needles made of bone, together with a greater variety of camping styles, were new to the European landscape. *Homo neanderthalensis* and their predecessors had been living successfully in this same region through many glacial cycles, but there is no evidence of even a single fish hook or sewing needle having been fashioned in Neanderthal times. Within 5,000 years of the arrival of "sapiens," the Neanderthals of Europe had vanished. A few fossils suggest limited gene exchange between the two species.[53] Recent DNA analyses confirm a bit of gene interchange between the muscular Neanderthals and their new neighbors.[54] Pale skin and reddish hair were likely Neanderthal adaptations, important to survival in seasonally frigid regions. (Paler skin is essential for producing vitamin D in one's skin when you're covered in animal skins under cold and cloudy skies. Dark skin protects against excessive UV irradiation and skin cancer in the sunny tropics.)

Because of their heavier skeleton and more *erectus*-like facial features, Neanderthal had been considered less intelligent and unable to speak by earlier scholars, despite having brains a bit larger than ours. But what were Neanderthals doing with such big brains if they weren't talking? Also, 300,000-year-old wooden spears found at the bottom of a lake in Central Germany, together with the butchered bones of wild horses, make clear that Neanderthals were effective hunters. Most likely the horses were killed and butchered on an ice-covered lake in winter, with spears and bones sinking to the bottom in springtime to become preserved under water. More significantly, these ancient Neanderthal spears lacked stone spear points; they created stone spear points later. It is not until around 40,000 years ago that sapiens-type stone spear points are first found in Northern Italy, linked with teeth having sapiens-type DNA. With advanced cultural skills, *Homo sapiens* grew to larger numbers, numbers that would replace the Neanderthals. Asian populations were also confronted by the slender "out of Africa" immigrants, and they too were transformed. Nevertheless, long distances have delayed genetic mixing; even today some New Guinea and Australian natives carry snippets of ancient DNA not found in other living humans.

Landscapes with limited resources, in which groups of humans came into competition, gave rise to progressive cultural dynamics. Within-group cooperation set the stage for between-group competition, throughout our range. Recent evidence from southern Africa, dated at 70,000 years ago, suggests that the modern mind was fully functional at that time. Here, spear points had been fashioned of three very different materials. A glue-like plant resin hardened with ochre and heated secured the stone point to the haft of the spear.[55] This skill required a shared tradition, dedicated apprenticeship, and careful execution. Such sophisticated weaponry is consistent with yet another fact: the human brain has not increased in size over the last 50,000 years.

> *Sophisticated technology, highly developed science and elaborate social or religious rituals are products of a cumulative process of cultural evolution, whereby each generation builds on the achievements of their predecessors in a gradual, approximately monotonic, ratcheting up of complexity and functionality.*
>
> —K. Smith et al.[56]

No longer confined to random mutations or genetic reorganization, cultural cooperation produced new behaviors, new technologies, and new possibilities. Most important, cultural inventions could be quickly adopted by succeeding generations or co-opted by competing societies. Here was a real-world example of "Lamarckian Evolution." The French evolutionist, Jean-Baptiste Lamarck, postulated that parents who became adept at a particular skill would beget children better able to carry on that same skill, but this doesn't occur in higher animals.[57] In contrast, cultural evolution does allow newly acquired skills to be passed on to succeeding generations.[58] Tool making was probably our earliest skill, quickly followed by food processing. Language—our premier talent— allowed us to share both timely and abstract concepts with our kin. Imitation, another fundamental human trait, was the basis of the "conservation and dissemination of innovations in ways that allow technologies and practices to improve over time."[59]

Though human numbers remained small, cultural evolution became *dynamically progressive and ecologically impactful.* Humans reached Australia over 50,000 years ago; by 40,000 years ago nearly all the large animals of that continent were gone! Spores of fungi specific to the dung of large herbivores decline rapidly about 45,000 years ago, followed by frequent fires and major changes in Australia's vegetation. This was not "climate change" but the impact of humans. Our species first entered the Americas from northern Asia around 15,000 years ago. Finely crafted stone spear points are first recorded in North America around 11,000 years ago, followed by the extermination of mammoths, mastodons, giant ground sloths, wild horses, and other nutritious treats. In 1910, A. R. Wallace suggested that the extinction of large mammals at the end of the ice ages were "actually due to man's agency." A few obtuse scholars still claim that "climate change" accounted for these extinctions, oblivious to the fact that these very same animals had survived many earlier glacial cycles. Clearly, our species, operating in small bands, had become the most widely distributed and dangerous mammalian species on Earth.

Then, quite suddenly, and beginning around 10,000 years ago, at least five human societies in several distant areas of the world initiated an extraordinary advance. We developed **agriculture** and **animal husbandry**, a kind of symbiosis that would grandly expand both our numbers and our potentialities! With cultural innovations, increasing complexity on planet Earth had now become purposeful: humankind was beginning to alter the planet!

Chapter 10

EVER MORE COMPLEXITY: HUMAN CULTURAL ADVANCE

Gaia is a thin spherical shell of matter that surrounds an incandescent interior; it begins where the crustal rock meets the magma of the Earth's hot interior, almost 100 miles below the surface.... I call Gaia a physiological system because it appears to have the unconscious goal of regulating the climate and chemistry at a comfortable state for life.

—James Lovelock[1]

Viewed from afar, Earth is a thing of beauty. White clouds swirl across the blue orb, bright deserts reflect warm sunshine, and green vegetation marks fertile landscapes. Not just beautiful, but geophysically, climatologically, and biologically interactive. Using *Gaia*—goddess of the Earth—as a metaphor, James Lovelock claimed that the geophysical planet and its biota were a mutually stabilizing, self-balancing, holistic entity. First framed at a time when ecologists found equilibria wherever they looked, the **Gaia Hypothesis** was enthusiastically embraced. That was back in the 1970s. Unfortunately, things look very different today.

The first data to shake the Gaia mentality came from Greenland's thick ice-cover; the second were deposits on the floor of the North Atlantic Ocean. Cores into both ice and sediment revealed a hundred thousand years in which sudden climatic shifts had taken place with some frequency. Not only had ice ages come and gone, but on some occasions climate had changed abruptly within a decade. *Gaia wasn't supposed to work that way!* Perhaps ocean currents, coursing across the North Atlantic, had suddenly changed direction. Then came another nasty surprise: atmospheric research revealed an expanding "ozone hole" high above Antarctica. Human-devised chlorinated cooling agents were disintegrating in the frigid stratosphere and destroying ozone in our upper atmosphere! And, as if all this weren't trouble enough, the world's been getting warmer!

Over a century ago, chemists had predicted that pumping ever more carbon dioxide into the atmosphere would warm the globe. Carbon dioxide has an unusual property: it is transparent to higher energy ultraviolet radiation but absorbs lower energy infrared energy. Ultraviolet energy from the Sun passes through the atmosphere, strikes the Earth, becomes converted to infrared and is re-radiated outward. It is this re-radiated energy that is partly absorbed by *greenhouse gases* and keeps our planet comfortable, rather like a blanket or the air within a greenhouse. Consequently, a higher concentration of CO_2 will cause greater planetary warming. Alaska's weather is changing especially fast. Inuit people of the far north are seeing robins for the first time; their language has no name for this distinctive bird. As carefully recorded over the last 1,200 years, Japanese cherry trees are now blossoming earlier than ever before. Our biosphere is getting warmer.

Gaia cannot cope! Though half our carbon dioxide emissions find a home in the ocean, soil, and plant life, the other half is being added to our atmosphere. Also, by producing agricultural fertilizers, humans are now fixing as much nitrogen as all the rest of the biosphere, and this is changing aquatic ecosystems. Having reached his nineties, Lovelock has lost his optimism and now sees Gaia as capricious. But how did this come to pass? How could humans have become so powerful a force, on so grand a planet, in so little time?

THE AGRICULTURAL REVOLUTION

> Homo sapiens *could never have become the highly technological, environment-dominating and environment-transforming animal of today if agriculture had not been invented.*
> —Paul Ehrlich and Anne Ehrlich[2]

Agriculture was truly something new and revolutionary. The balance between humans and the rest of the biosphere was decisively altered.

—Jeffrey Sachs[3]

Humankind's most significant recent advance may have been a natural outcome of increasing cultural sophistication in response to mankind's ever-present fear: **starvation**. Though hunter-gatherers had survived the ice ages in both tropical and temperate environments, our numbers remained small. As with every other species, human survival was precarious. Local populations were regularly challenged by prolonged drought, unusual cold, or virulent pathogens. Finding enough food to survive and reproduce was a never-ending challenge. Then suddenly—within a geological instant—our species diminished its likelihood of starvation by initiating a series of new symbiotic relationships! We call these innovations **agriculture** and **animal husbandry**.[4]

By nurturing a few carefully chosen species of plants and animals, humans assured themselves a more reliable food supply. Herding sheep and goats may have come easily for nomadic hunters. Capturing young animals and caring for them afforded a ready supply of meat and hides. Since these animals lived naturally in small social groupings, they adapted well to their new masters. Soon we were corralling, breeding, and herding them. Best of all, these hoofed animals eat plant materials we humans cannot digest, allowing more of our number to prosper in grasslands and thorn bush. In different parts of the world, sheep, goats, cattle, donkeys, horses, camels, yaks, and llamas became essential members of human communities. Chickens in southern Asia and turkeys in Mexico added to the feast. Though dogs had been our hunting companions for perhaps 20,000 years, both dogs and pigs, feeding on village refuse, became part of the menu in many cultures.

Learning to plant and tend selected plants may have been more challenging than herding sheep and goats. Finding nutritious plants amenable to cultivation, planting and weeding them, carefully harvesting, then separating and storing seeds for next year's planting, were sophisticated new activities. Perhaps most extraordinary of all, these life-transforming advances were initiated by people in several corners of the globe *independently!* Agriculture in the New World arose totally apart from that practiced in the Old World. North America's first settlers were pre-agricultural hunting bands. By three thousand years ago, Mexican and Mesoamerican farmers were growing maize, squash, tomatoes, cacao, avocado, papaya, and vanilla, as well as local varieties of chili peppers, kidney beans, and cotton. People in South America, from the high Andes to the Amazon lowlands, first domesticated potatoes, cassava, sweet potatoes, pineapples, peanuts, and tobacco. They, too, came to grow distinctive varieties of maize, beans, chili peppers, and cotton.

In the Middle East, agriculture appears to have arisen earlier than in the New World. Here, people were nourished by durum wheat, barley, bread wheat, chickpeas, lentils, olives, grapes, dates, figs, and many other foods. People first used bananas, several kinds of melons, eggplant, mangos, and many other fruits and vegetables in southern Asia. Several varieties of rice and millets, as well as soybeans, were first developed in eastern Asia. Inhabitants of New Guinea and the western Pacific were the first to harvest sugar cane, local yams, taro, and coconut palm. Western and northern Africa are the original home of sorghum, yams, several millets, indigenous melons, okra, oil palm, native cotton, and cola seeds. Highland Ethiopia is the home of the world's smallest cereal grain (*teff*), a banana relative used for starch (*ensete*), a useful oil (*noog*), a stimulating beverage (coffee), and distinctive varieties of both sorghum and barley.

Around a thousand years ago, people in eastern North America were cultivating native goosefoot and marsh elder, as well as new varieties of sunflowers, pumpkins, and maize. Among all these many plants grown around the world, grains (from the grass family) and pulses (from the legume family) became widely used staples. Other cultures used root crops as their primary starch source, such as potatoes in the high Andes and cassava in the Amazon basin. Because they lived on the same super-continent and along similar latitudes, cultures in Asia and Europe easily shared crops and domestic animals. Overall, agriculture was initiated between ten thousand and five thousand years ago in several regions of the world. More to the point, agriculture allowed us to store nourishment for bad times, becoming the most successful species the world had ever seen![5]

While we still do not understand why so critical an advance in human history occurred at this particular period in time, the results have been transformative: *agriculture fed more people more reliably!* After the introduction of agriculture, habitation sites in Western Europe expanded ten-fold in area, implying a ten-fold increase in local population! That was the good news, but agriculture had downsides as well. Living a sedentary life in close contact with domesticated animals fostered more diseases. In crowded settlements, people often suffered nutritional deficiencies, becoming smaller than their nomadic antecedents. Nevertheless, these negative effects had little impact. Agriculture allowed humans to beget more humans. Called the **Neolithic Demographic Transition** by historians, our increasing numbers advanced the elaboration of human societies.

Viewed from the perspective of the Red Queen, humans were beginning to run a lot faster than anyone else. Agriculture had boldly increased our species' *fitness*. A world abundant in biodiversity provided the resources on which this advance was constructed. From amongst more than 260,000 described species of flowering plants, twenty-five species provide more than 85 percent of our daily energy requirements! (This includes the plants that nourish the animals we eat). A few thousand additional flowering plant species give us important spices, oils, condiments, fibers, and medicines. Without this grand diversity, we could not have provisioned so grand an expansion of human numbers. And let's not forget the soil. Farming depends on soils that have become organically enriched over many millions of years. Similarly, coal, oil, and gas are all the products of earlier life activities, empowering today's industrial societies.

FEEDING THE MULTITUDES

Agriculture allowed larger numbers of people to live together in stable settled communities. In turn, larger sedentary human communities allowed for both stratification and specialization. Men as hunters and warriors, and women as nurturing mothers and gatherers, had been playing their essential roles for thousands of generations. But now, with agriculture and increased numbers, larger human societies began to behave like "superorganisms." Resembling the castes found within insect colonies, one human caste could be devoted to food production, others to working wood or metal, and others waging war. Even so, biology determined destiny. Females were absolutely critical in birthing and rearing the young; our large brains required constant nurturing. Males, with greater upper-body strength, tilled the soil, worked metal, constructed buildings, and defended the community.

Not only did larger agricultural populations expand human demands and intensify human competition, they provided a platform for further rapid and far-reaching innovation. As Geerat Vermeij has written, "Competition in the broadest sense . . . is universal in all economic systems, including life itself, at every scale of inclusion."[6] Regardless of whether the economy was fueled by photosynthesis in natural ecosystems or driven by the territorial imperatives of feuding human polities, competition within and between societies has made us who we are. It is our species that now stands at the pinnacle of earthly dominion. Our croplands and pasturelands are today's largest terrestrial ecosystems.

Thanks to agriculture, forestry, aquaculture, animal husbandry, fishing, and other activities, we are currently devouring over 30 percent of the Earth's net primary productivity.[7] Agriculture was the foundation on which the further advance of industrially powered human societies was constructed. Humans and their "agricultural symbionts" have become one of the most transformative symbioses in the history of life. We propagate and carefully tend rice, sheep, pigs, potatoes, and lots more; then we eat them!

Settled villages, supported by agriculture, provided humans a launching pad for a new trajectory—accelerated cultural and technological advance. While nomadic hunters may have been efficient in survival and added new information to ancestral lore, their numbers remained small.

Suddenly, along major rivers and with carefully engineered irrigation, agriculture could support thousands of our kind. Strategically located villages blossomed into city-states, providing a new platform for human advancement. Protected within the city's walls, people were free to utilize their many talents and fashion innovations.

Ovens for baking bread and firing useful ceramics were soon followed by forges working metal. Early copper work was followed by an age of bronze in the Middle East. Here again, our planet played a critical supporting role. Three tectonic plates abut in the Middle East. This is where minerals—molten at great depths—became concentrated and returned to the surface, explaining why metallurgy began so early in this corner of the world. (The island of Cyprus is named for copper.) Then, using hematite as a flux in working copper, and perhaps accidently, metal workers of the Middle East began extracting iron from a reddish rock. High temperatures under low oxygen levels allowed charcoal to remove oxygen from hematite, yielding metallic iron, and this, with a dash of carbon, gave us steel. Farmers and warriors demanded stronger, harder, sharper metals. Steel tools felled forests more quickly, plowed the ground more effectively, and waged war more ruthlessly. Requiring high and precise temperatures, **metallurgy** was another singular advance in our species' advance.

Cities promoted the division of labor and technological innovation. Planting crops in seasonal environments required understanding the Sun's annual cycle: the beginnings of astronomy. A wide variety of wild plants provided medications; their proper identification and careful application fostered the rise of medicine. Agricultural transactions demanded accounting; writing followed quickly. Easily stored and transported, cereal grains and pulses fed the masses and fueled a mobile military. Guarding transportation routes, imposing order, and fending off attacks, armies promoted commerce while securing empires.[8] Stone tools had changed little over thousands of generations during our early history, but now, with settled agricultural societies, human technologies advanced rapidly.

Much earlier, and as the only species on the planet to be fearful of the future, humans had developed mythic traditions. Religious belief infused our lives with hope and meaning, strengthening our resolve in the face of uncertainty and calamity. In urban centers, religious expression fostered unity, supported a ruling elite, and inspired monumental architecture. Religious values validated jurisprudence, providing order in crowded societies.[9] Claiming divine sanction, rulers promised justice and stability. All the while, diverse human talents found new opportunities within the city-state. Prodded by wealth and warfare, urban people conformed their lives, advanced their skills, and prospered like never before.

Cultural advance, however, has differed greatly around the globe. Native Australians and other isolated groups remained hunter-gatherers. Native Americans developed a rich agricultural tradition and created complex societies with impressive monumental architecture; however, they had no strong draft animals, and lacked iron and steel. In contrast, the people of Europe, North Africa, and Asia shared a vast land area replete with rich and varied resources, Novel discoveries and continuous interaction allowed Eurasian societies to become richer and more complex.[10]

Social norms influenced progress as well. Around 1450 CE, China chose condescending isolation to more dynamic interaction, even as Europeans were beginning to explore the world. Mediterranean cultures proved more adventurous, perhaps as a consequence of active trading and incessant warfare. By 600 BCE, state-supported currency had accelerated Mediterranean trading. Lacking a strong religious priesthood, classic Greece was free to foster philosophy and science, even experimenting with democracy. Later, Rome excelled in law and civic engineering, managing a grand empire over five hundred years. Two hundred years after Rome's collapse, Islamic societies fashioned a trading network linking Morocco to India and China. Having taken control of Samarkand, the Islamic world was introduced to Chinese papermaking, the perfect medium for recording and translating earlier knowledge into the language of the Koran. Between about 800 and 1100 CE, with new mathematical techniques, adopting a numbering system from India and studying both vision and astronomy, scholars writing in Arabic grandly advanced the sciences. Tragically, with the destruction of Baghdad's libraries by Mongol invaders in 1258, the Islamic efflorescence wilted. Around the same time, with knowledge translated from Arabic and Greek into Latin, scholarship became revitalized in Western Europe.

A SCIENTIFIC REVOLUTION

> *Looking at the history of science, we do find a net increase in the empirical success of our theories and unquestionably also an increase in the degree to which they constitute a unified, integrated, mutually supported network.*
>
> —Keith Parsons[11]

Blessed with good weather from the tenth through early fourteenth centuries, Europe grew in population and prosperity. Divided by mountain chains, and with many regional principalities, Europe could not be bound into a single empire as was China. Sharing a single language— Latin—for both religious and intellectual discourse, European culture proved open and innovative. Beginning with a compendium of Roman law, Medieval Europe created a pragmatic new civil legal system, marginalizing religious jurisprudence and fostering a more dynamic mercantile society.[12] Because of its many navigable rivers and mineral resources, Europe's growth shifted northward, away from the Mediterranean. Challenged by newly translated Arabic and Greek texts, European scholars began their own intellectual journey. Claiming that a closer study of nature would illuminate our understanding of God, Medieval Christian philosophers gave social sanction to the careful study of God's creation. Shifting away from the certainties of traditional knowledge, scholars now confronted what we did not understand. In this way, Western Europe began an intense examination of the natural world.

Calamity briefly intervened. With increasing trade from the east, Europe was assaulted by a virulent pathogen. Beginning in 1347, the Black Death nearly halved the population of Europe. But again, and as elsewhere in the history of life, calamity afforded new opportunities. Following Europe's pandemic, individual labor became more valuable, land became less expensive, and a Renaissance soon blossomed. All the while, marine transport along both Atlantic and Mediterranean shores continued apace. The stern-post rudder, compass, and gunpowder (all from China) accelerated naval activity. Prosperity sharpened Europe's appetite for tropical spices, fine ceramics, and Chinese silks. Satisfying these demands would create wealth for trader, merchant and banker.

To this purpose, Prince Henry of Portugal initiated a program of naval exploration southward along the western coast of Africa. Sturdy little ships, caravels were built to explore unknown shores and return with knowledge for further voyages. These many expeditions achieved their goal in 1497, when Vasco da Gama sailed southwest across the Atlantic toward Brazil, turned eastward to catch the prevailing winds, and circumnavigated Africa to reach Calicut, India. Portugal's naval exploration had linked Portugal to India, providing Europe with precious goods at lower costs. Indian Ocean navies confronted these invaders, but the Portuguese had cannon balls to spare. Earlier, in 1492, certain that he could reach Asia more directly, Columbus sailed directly westward across the Atlantic. Instead of Asia, he collided with a "New World," making clear how much was yet to be discovered!

Europe's own inventions, such as iron horseshoes, eye glasses, mechanical timepieces, and double-entry accounting, led to greater efficiencies in agriculture and commerce. More fundamentally, Europe had a free and open market system. Legal agreements and interchangeable currencies provided a *social technology* in which commerce would flourish. Paper-making reached Italy in the late 1300s. This useful medium allowed Johan Gutenberg to print books using precisely fashioned interchangeable metal type. Costly manuscripts, laboriously hand copied in times past, were suddenly affordable in printed form. Printers and their presses (as venture capitalists) moved across Europe, grandly expanding the sharing of knowledge. In stark contrast, both China and the Islamic world shunned Gutenberg's technology for almost two centuries. With affordable Bibles and his own vernacular translations, Martin Luther urged Christians to study the scriptures for themselves. Challenging Vatican authority in this way, Europe became convulsed in religious and political turmoil, even as commercial activity and scientific inquiry expanded.

Around 1608, a Dutch optician devised the first telescope, allowing him to see farther more clearly. Learning of this invention, Galileo constructed his own in 1610, watching as four moons circled Jupiter over the following weeks. Viewing Venus over several months, Galileo witnessed the phases of that planet as it journeyed around the Sun, exactly as Copernicus had proposed! About the same time, and using Tycho Brahe's carefully recorded planetary observations, Johannes Kepler deduced the rules of planetary motion. Considering this astronomical data from the perspective of a falling apple, Isaac Newton explained planetary motion with the elegant concept of universal gravitation in 1687. Exactly as natural philosophers had claimed, careful observation revealed God's universe was governed by laws! The scientific revolution had begun.

AN INDUSTRIAL REVOLUTION

> *Industrialization made possible immense improvements in human*
> *health and longevity, while providing ordinary people with greater*
> *material welfare than their ancestors could ever have imagined.*
> —Patrick Alitt[13]

Growing in population, enmeshed in constant warfare, and demanding goods from all around the world, Europeans were consuming ever more resources. By the late 1600s, the British Isles were suffering a timber famine. Warships, general construction, and the need for firewood had devoured forest and woodland.[14] Coal proved to be an efficient replacement for heating needs, but coal mines flooded frequently in rainy Britain. Steam engines, developed in the mid-1700s to pump water out of the flooding mines, *converted heat into work.* By the early 1800s, more efficient steam engines were propelling cargo along rails of steel. Railroads were a new kind of technology—rail lines, engines, cars, and stations in a linked network. Factories with steam-powered looms made cotton textiles more affordable. Also, as the economy grew, Europeans craved more sugary sweets. Intense demand for both sugar and cotton gave rise to a harsh industrial slavery in the Americas, even as European peasantry was conscripted into factory labor. All the while, modern capitalism accelerated economic advance by investing surplus profits into further production and innovation.

Coal-powered steam engines had begun another grand advance in the human saga. No longer limited to human, animal, wind, or water power, the **Industrial Revolution** ran on fossil fuel. Powered by this energy-dense legacy of ancient life, both human numbers and human technologies expanded at an ever-quickening pace.[15] Scientific investigation and experimentation, publicly vetted and subject to verification, produced extraordinary dividends. In 1821, Michael Faraday showed that a rotating magnet generated electric current in a copper wire. A few decades later, engineers were transmitting electric power over long distances. Using small confined explosions, the reciprocating internal combustion engine proved both efficient and mobile. In 1915, one-fifth of America's farmland was devoted to feeding horses; they were soon replaced by gasoline-burning tractors. Having seen Chicago butchers process beef on overhead conveyers, Henry Ford assembled automobiles more rapidly with overhead assembly lines. In turn, affordable automotive transport accelerated commerce and expanded personal freedom. Heavier-than-air flight, first by propeller, then by jet, allowed us to traverse the planet, transporting both people and precious goods quickly over large distances. Today, consuming large stores of energy-dense fossil fuel, we enjoy lifestyles unimagined in earlier times.

Energy consumption has been central both to the evolving odyssey of life and to our ever-expanding human economies. For over two billion years, oxygenic photosynthesis provided the living world its motive force. Concentrated over many millions of years, ancient life has given us fossil fuels rich in chemical energy. Beginning the Industrial Revolution with coal, fossil fuels expanded technological development with oil and gas, and now support lifestyles of ever-greater comfort. Heated in winter, cooled in summer, with refrigerated foods throughout the year and jetting to destinations around the world, we humans are consuming more and more fossil energy. Modern society and its many technologies mark the most recent chapter in Earth's ever-increasing complexity. Biological evolution, working slowly over millions of years—powered by the Sun—resulted in greater biodiversity and more elaborate ecosystems. More recently, we humans discovered new ways of improving our lives, at first with simple hunting tools and fire, later with agricultural crops and domesticated animals. Today, modern societies are energized with fossil fuels, atomic fission, water power, wind, and sunshine. All the while we mine rich ores concentrated over billions of years by the Earth's dynamic geology. Empowered in these various ways, we are elaborating our technologies more rapidly. But before we engage the consequences of our technological advance in our final chapter, let's take a quick look back over Earth's long and episodic history of ever-increasing complexity.

Chapter 11

A FOUR-BILLION-YEAR EPIC

During his acclaimed television series, *The Ascent of Man*, Jacob Bronowski took a moment to look directly into the camera and ask viewers what the probability might be of chemical elements coming together randomly to form a single human being? His point was clear: such a probability is *zero!* The likelihood of random chemistry coming together to form a weed or even a bacterium is similarly impossible. Bronowski's point was simple: complex life forms are the result of developmental programs that came into being and were slowly elaborated over more than three billion years. Innovations were initiated during particular times and became the template for further modification over time.

For evolutionary biologists, the concept of innovation has been contentious over many generations. Does innovation arise from the gradual accumulation of small steps? Or does it depend on the "emergence" of new traits, suddenly becoming the "platform" for new trajectories? Obviously, small changes in base-pair sequence are likely to yield small changes, but the accidental duplication of an entire gene may open more significant possibilities. Surely, both have contributed to evolutionary innovations; they have an historical past, arose within a period of time, and may have determined the future in extraordinary ways. Modern human societies have, in fact, become one of the most extraordinary innovations in the history of life. We ourselves, and all other living things, are the product of a very long history. The record in the rocks, our own development from a single cell, and the genes within our bodies, all bear testimony to a very long epic.[1] Let's review this grand trajectory of life's elaboration—and human achievement—by dividing it into ten major stages.

1: LIFE'S ORIGIN

Surely, the first major step in our story was the origin of simple life itself. Having left no trace, the earliest attempts at biochemical organization and biophysical energetics were likely devoured by their successors. Protein synthesis within a protective vesicle allowed for metabolism. RNA conveyed instructions within the cell, while DNA carried information across generations. Today, bacterial-grade cells are the simplest forms of life. Able to gather and utilize energy, to grow and replicate, bacteria can transfer the information for all their life activities from generation to generation. These are the hallmarks of life.

Today, life arises only from living cells already formed. All your membranes, mitochondria, and cytoplasmic architecture were derived from your mother's egg cell. (Dad's sperm cell added a set of chromosomes to begin your existence, but little more). You began as a single cell, as does all other life! Only two aspects of living cells are never formed *de novo*: cell membranes and the DNA helix—all the other stuff can be built anew. How then did life begin?

A very unusual aspect of the living world is that, while DNA is life's fundamental information carrier, Bacteria build their DNA in a fundamentally different way than do Archaea and Eukaryota. Add this to Eukaryota having Bacteria-like operational genes and Archaea-like informational genes, and you understand that some biologists believe that there may never have been a "first living cell." Instead, early life may have arisen from an organic ecosystem: "a community of quasi-cellular entities that exchanged genes and evolved collectively and from which the three domains of life crystallized."[2] In fact, early viruses may have shuttled genes around between these early life forms in the vicinity of deep sea alkaline vents, in temperatures well below the boiling point of water.

Some scholars have claimed that organic chemistry can exhibit *inherent self-organizing properties*. These properties, they argue, gave rise to replicating structures spontaneously. Indeed, proteins fold themselves into complex shapes automatically, based on how specific amino acids are arranged along their length. All that was required was a stable aquatic environment, organic molecules cooperating together within enclosing vesicles, various energy sources to power replication, followed by millions of years for trial and error. Indeed, given the proper conditions and sufficient time, life's origin on planet Earth may have been inevitable.[3]

2: CAPTURING THE SUN'S ENERGY

With cellular organization and reproduction in place, bacteria could use a variety of chemical reactions to keep themselves going. But metabolic reactions were restricted to special environments with energy-yielding substrates. The second major advance in our grand epic was capturing the energy of sunlight to build complex organic molecules. Purple photosynthesizing bacteria were probably the first to achieve this breakthrough. Using the energy of sunshine, they acquired hydrogen from a variety of sulfur compounds. Unfortunately, hydrogen-yielding sulfur compounds can be hard to find. Oxygenic-photosynthesis overcame this limitation by acquiring hydrogen directly from the water molecule. Here was a way of obtaining energy-packed hydrogen atoms just about anywhere on the sunlit Earth! By uniting hydrogen—*torn loose from water*—with carbon dioxide, **oxygenic photosynthesis** by Cyanobacteria transformed the energy of sunlight into energy-rich carbohydrates, while releasing free oxygen to the atmosphere. Chlorophyll and its associated pigments are the central players in this dance of sunlight, physics, and chemistry.

Once life on Earth had learned how to rip apart the water molecule and build energy-rich foods, organic evolution began its grand progression. Ignoring the second law of thermodynamics—where everything runs down—oxygenic photosynthesis provided the energy needed to run the complex dynamics of metabolism and other life activities.[4] Over time, this same energy allowed things to get more complicated. That was the good news; the bad news was that Cyanobacteria were polluting the atmosphere with a very nasty poison—free oxygen. Highly reactive, oxygen is a lethal threat to many forms of life. (Oxygen-averse bacteria are still with us today, active only in oxygen-poor niches.) But calamity for some was opportunity for others, as a new lineage of bacteria invented **oxygen-consuming respiration.** By tearing carbon dioxide out of a simple carbohydrate, and uniting the remaining hydrogen with oxygen, respiration provided more energy more efficiently.

3: A LARGER MORE COMPLEX CELL

The third major stage in the history of life was the development of the larger eukaryotic cell. Characterized by having a nucleus (within which the chromosomes are protected and replicated), these cells are a thousand times more ample in volume than the average bacterium. Central to their success was acquiring mitochondria, bacteria-size organelles in which respiration breaks down carbohydrates to provide energy for the larger cell. Oxygen-consuming respiration yields over ten times more energy than is gained from the fermentation of simple sugars. With their efficient little mitochondrial fuel-cells, eukaryotic cells were able to process more information and become more complex.[5]

Meiosis and the production of sex cells allowed for both the repair of genetic information and a lottery of new allelic combinations as the union of egg and sperm produced a new generation. Sex provided for a continual remixing of hereditary traits, facilitating adaptation to changing environments and evolving pathogens. Eukaryotic cells were the platform from which life's complexity advanced.

Similarly, the incorporation of yet another bacterial partner within a eukaryotic lineage marked another major advance. This new partner was a cyanobacterium, carrying with it the power of photosynthesis. These new endosymbionts became transformed into chloroplasts, allowing green algae to become photosynthesizers themselves. Red algae, brown algae, and several other lineages acquired additional photobionts as well. Algae of all kinds, together with the activities of the Cyanobacteria, expanded oxygen-producing photosynthesis around the world. Slowly, our atmosphere became breathable.

4: MULTICELLULAR LIFE

The fourth major stage in our continuing epic of progressive change was building larger, multicelled, organisms. Originating independently in three eukaryotic lineages, this new process gave us the three great kingdoms of larger living things: plants, fungi, and animals. Each of these kingdoms constructs its multicellular forms in entirely different ways. With cell walls made of thin membranes, animal embryos can undergo complex development, involving both cell migration and self-dissolution. Fungi build their larger forms, whether toadstools or lichens, with hair-thin hyphal threads, elaborately woven together. In contrast, algae and land plants have cells whose walls are built of stiff cellulose. One of the world's most complex polymers, cellulose builds brick-like cells, strong but static, which is why plants find it so difficult to move.

All multicellular life forms, regardless of the kingdom, require cell-to-cell adhesion and sophisticated intercellular communication. Systems of intercommunication are essential both for growth and structural integrity. In fact, all multicellular beings are constructed of eukaryotic cells. Apparently, the greater information-carrying-capacity of the eukaryotic cell was a necessary prerequisite for the construction of complex multicellular life. Without the mitochondria-powered eukaryotic cell—and an atmosphere rich in free oxygen—the world could not have produced larger, more complex, multicellular beings.

5: ANIMALS ARRIVE

By my count, the fifth major stage in the advance of life's complexity was the development of larger and more active animals during what has been called the Cambrian explosion. As we discussed in chapter 7, around 540 million years ago (mya), the world's oceans suddenly became home to a zoo of different animal lineages! Worm-like animals, shell-bearers, and many-legged crawlers all appear together at about the same moment in time. Many of these creatures developed eyes to see where they were going, and chewing mouthparts to devour one another. The paucity of trace fossils—such as worm trails and foot-prints—makes clear that animals were miniscule in earlier times. An atmosphere with increasing **oxygen pressure** was likely the key factor, allowing respiration to power both larger creatures and more active pursuit in the animal world.

Though oxygen-consuming respiration may have powered larger, more complex, animals, it didn't build them. Increasing complexity required more elaborate developmental protocols. Over time, accidental gene duplication provided the wherewithal for new developmental pathways. These more varied instructions produced an increasingly diverse fauna. Preceded in the fossil record by the enigmatic Ediacarans, who then declined, the sudden appearance of Cambrian animals marked another grand advance in the history of life.

6: PLANTS PIONEER THE LAND

The sixth major advance in the history of life, and surely one of the most important for the diversity of the living world, was the colonization of land by green plants. With embryonic tissues at their growing tips, land plants elaborate many forms, from weeds to trees, and spores to flowers. Once land plants had acquired lignin-strengthened cell walls and a more efficient plumbing system, they grew larger and more complex. With a tubular cambium, expanding in strength and fluid transport every growing season, woody trees formed tall forests. Land plants added organic matter to the soil, reduced erosion, provided moisture and shade to those within the understory, and nourished a terrestrial fauna.

Another singular botanical advance came with the development of ovules and pollen, giving rise to seeds. Carried by the wind or animal vectors, pollen grains germinate near the ovule to effect fertilization. Pollination supplanted swimming sperm, allowing seed plants to reproduce in drier habitats and create a broader variety of biomes. As plants pioneered land surfaces, a number of animal lineages began their own terrestrial diversification. Of these, one has become preeminent.

7: VERTEBRATES CONQUER THE LAND

From a human perspective, the seventh major step was the emergence of land vertebrates. Similar to land plants and insects, four-legged vertebrates arose from fresh-water environments, beginning their advance onto the land about 370 mya. Here was a lineage of animals that could grow large, defy gravity, move quickly, and build more complex brains. A terrestrial flora, supporting insects, spiders, worms, and snails, made it possible for early land vertebrates to find food. Ever since their origin, vertebrates have been the largest terrestrial animals. We humans bear uncomfortable witness to the rise of land vertebrates, choking readily because air and food passageways are not clearly separate in our throats—just as in our fish-like ancestors. Gill-depressions that come and go during early human embryonic development are further evidence that we arose from fish. Four limbs and a many-parted backbone are additional testimony to our tetrapod vertebrate heritage.

Though life rebounded after the massive Permo-Triassic extinction 250 mya, biodiversity was clobbered yet again with the end-Cretaceous extinction event 65 mya. This last extinction, with ongoing volcanic activity punctuated by an asteroidal blast, obliterated the most charismatic lineage of land animals: the dinosaurs. Dominating the world's landscapes for over 200 million years, only birds remain as living descendants of these extraordinary animals. That was the bad news; the really good news was that smaller furry creatures suddenly found themselves in a less dangerous world. After the departure of the dinosaurs, and already represented by a number of lineages, furry mammals diversified explosively. Fortunately, and in advance of the end-Cretaceous extinction, Mother Nature had fashioned yet another grand advance in life's diversity.

8: FLOWERS, FRUITS, AND GRASS

The eighth grand advance, I'm convinced, was the development of the flowering plants (Angiosperms). Ranging from little floating pond weeds to orchids and baobab trees, flowering plants come in an extraordinary variety of shapes and sizes. Whether counted by species number or standing biomass, they make up the vast majority of terrestrial vegetation, providing both architectural variety and nutritious abundance. Of their many clever adaptations, getting animals to carry pollen grains from one colorful flower to another was a significant advance. Not only did the flowering plants use animals as pollinators, offering them sugary nectar, they also supported many animals with a grand variety of yummy fruits and nutritious seeds. Having enclosed their seeds in a special encasement, flowering plants were able to fashion everything from the little grass grain to the soybean, tomato, pumpkin, and coconut. Flowering plants, together with their animal pollinators and fruit dispersers, have grandly expanded biodiversity over these last one hundred million years.

Another recent innovation has also been transformative, both in drier ecosystems and for human history itself. This was the expansion of grasslands. While grasses may have existed before dinosaurs were extinguished, they have become important only in more recent times. Antelope, deer, cattle, and horses all increased in numbers as grasslands have expanded around the world over the last thirty million years. Because their slender stems grow from a tufted and protected base, grass survives both fire and grazing. With the spread of C_4 grasses about six million years ago, fire-prone grasslands have expanded their dominion in warmer and drier environments.

9. A TWO-LEGGED PRIMATE GETS REALLY SMART

As the pageant of life continued, flowering trees inspired the evolution of monkeys and, later, fostered the evolution of swinging apes. Stereoscopic vision and social lives had given primates larger brains. With more erect bodies, flatter chests, broader shoulders, and highly versatile forelimbs, the swinging apes became brainier yet. No other group of animals can swing their arms around the way we Homonoids do. Around the same time, expanding grasslands in seasonally dry habitats promoted the proliferation of large mammalian herbivores. These four-legged nutrient-rich resources, in turn, allowed a two-legged African ape to triple its brain volume in only three million years. Recall the TV advertisement in which the old lady asks, "Where's the beef?" The answer from an ecological view is simple: "On the grasslands and savannas!" There isn't much beef-biomass in dense evergreen forests; instead, broad open grasslands support extensive populations of grazers. Though omnivorous, our ancestors knew that large grazing animals provided food of the highest quality. Moving on two legs allowed our ancestors to throw rocks, fashion spears, and become effective hunters.

Rather like small colonies of ants, human bands began behaving as "super cooperators," competing with other human groups for access to food and water. Group bias, ethnocentrism, and the grand divide between "us-and-them" are the consequence of our evolutionary past. We are a tribal species whose primary environmental antagonists were other tribes. Surely this is the dynamic that drove brain expansion in close synchrony with the elaboration of speech. Rapid vocal communication empowered a cooperative coalition of males to protect their territory and resources. Inter-group competition not only expanded the human brain, it drove us out of the tropics into ever-more challenging environments. By carefully fashioning tools and weapons, cooking food to make it more digestible, and with language springing from our lips, humans became the most versatile animals to ever walk the Earth.

I disagree with those scholars who have imagined a sudden advance in human intelligence around 100,000 years ago. Yuval Harari crystalizes this notion as a "Cognitive Revolution."[6] Expanding three-fold in three million years, I suspect our intelligence expanded slowly over that time period, refining our speech, expanding our imagination, and giving us ever-more sophisticated social interactions. The expansion of *sapiens* populations over the last 200,000 years was, I suspect, a cultural phenomenon; we already had the brains.

10: HUMAN CULTURE EVOLVES PROGRESSIVELY

Three major advances have propelled human dominion ever forward. The first was becoming **bipedal** and erect, freeing our front limbs to carry, craft, and gesture. No longer needed as support, our hands became more flexible and more versatile. A second step was melding hand signals with our voices to create **language**, allowing for the rapid exchange of information and shared decision making. Tribal traditions became subject to competition and selection, propelling cultural evolution even more rapidly. The third grand advance in the human trajectory was developing a series of intimate symbioses with a variety of nutritious plants and a few tasty animals. **Agriculture**, like language, was a social technology that would increase our numbers and advance our dominion. Being able to store food for lean years, humans now flourished in larger settled villages, allowing yet more elaborate cultural innovations. Just as in large insect societies, larger city-states developed specialized castes to perform defined duties; human societies became highly cooperative *superorganisms*. Living on an older planet, plate tectonics provided concentrated ores, while erosion had amassed deposits of iron ore, coal, oil, and gas. These resources were the foundation for another grand advance—metallurgy. Ever-growing knowledge, together with a more critical analysis of nature, gave rise to the scientific revolution. More recently, we've used fossil fuels to power an industrial revolution, increasingly complex technologies, and ever-more comfortable lives.

COOPERATION AND COMPLEXITY

What are the factors that have given our planet increasingly complex biotas and humans ever-more creative cultures? Energy was the fundamental driver. With oxygenic photosynthesis capturing the energy of sunshine, the living world had a source of continual power. But complex living things break down over time and must reproduce themselves continually to survive in a difficult world. Since reproduction cannot be 100 percent accurate, the replicative process suffers inevitable mutations. These variants, subject to **natural selection** in changeable environments, have given us an ever-evolving epic of life. Fortunately, there was yet another significant factor.

In a remarkably insightful book, *Super Cooperators*, mathematical biologist Martin Nowak claims that **cooperation** "is the third principle force driving progressive evolution" after **mutation** and **selection**! In earliest times, he argues, molecular systems that cooperated to gather energy and stabilize their structure prevailed over their non-cooperating neighbors—becoming the "pre-life" from which replicating cells emerged.[7] The larger eukaryotic cell, within which a variety of organelles worked in a cooperative manner, proved to be a major advance. These more complex cells gave rise to multicellular organisms, within which thousands to trillions of cells work in tightly regulated unison. Finally, a few animal species have taken **social cooperation** to a level where a society of individuals acts as if were a single integrated assemblage: a *superorganism*. Army ants, honeybee colonies, and human societies are examples.

Comparing various cooperative versus selfish strategies in extended computer simulations, Nowak elucidates the origin and maintenance of altruism within both insect and human societies. He argues that cooperation was foundational to our premier talent: language. The advantage of better communication between individuals was especially significant as they challenged other groups for access to limited resources. Nowak's studies make clear that within-group cooperation in an environment of inter-group competition has been central to making us the dominant species on the planet. Today, cooperation is epitomized by multinational teams of scientists exploring subjects ranging from the Higg's Boson to cancer and cosmology. Working together to produce more food and confront our diseases, we have propelled our numbers into the billions.

THE HOLOCENE HAS BECOME THE ANTHROPOCENE

Empirically, we humans have extended ourselves across and into every ecological niche on the planet, making it impossible to say anymore where humans end and nature begins.

—Paul Wapner[8]

Over the last 2.6 million years, planet Earth witnessed a series of ice ages geologists defined as the **Pleistocene** epoch. This was arbitrarily terminated by the beginning of the **Holocene**, about 11,700 years ago. Nowadays, scholars of many persuasions are using the name Anthropocene in place of Holocene. And why not? We "anthropoids" have created a new epoch! In fact, 11,700 years ago is close to the time when elegant stone spear points—the Clovis tradition—first appeared in the archaeological record of North America; this is exactly the time when many of the New World's larger mammals began to disappear. Surely, this is reason enough to dump the Holocene and replace it with **Anthropocene**.

Nevertheless, human numbers did not expand significantly until we initiated animal and plant domestication, beginning 10,000 years ago. Alongside reliable rivers, hydraulic agriculture could now support thousands of our kind. With stable agricultural communities in place, human cultural creativity found new ways for us to expand our knowledge and our skills. Greek civilization promoted literature, science, and mathematics. Islamic scholars gathered, translated, and expanded scientific understanding. From India came a numbering system that empowered mathematics; from China came the compass, gunpowder, paper, and printing. Spurred on by aggressive mercantilism, inspired by an evangelical religion, and empowered by advanced weaponry, Europeans began a global hegemony. Coupling scientific inquiry and technological innovation with the latent energy of fossil fuels, the scientific and industrial revolutions have accelerated technological elaboration ever more rapidly. "Consecutively progressive," both science and technology have allowed us to better understand the world in which we live, constrain our diseases, nourish ourselves more generously, and increase our numbers explosively.

Various scholars begin the "Anthropocene" with agricultural expansion or early civilizations in Egypt and Mesopotamia. Others begin with the Industrial Revolution, or the use of atomic bombs in Japan. In her book *Adventures in the Anthropocene*, Gaia Vince begins the Anthropocene with the end of World War II, using the Holocene as a contrast to the extraordinary changes that have afflicted our home planet over these last seventy years. Replete with extensive research and a wealth of data, the author is optimistic, writing, "The Anthropocene could become a time of more nuanced climate change, where temperature and precipitation are modulated to humanity's needs. Where weather is planned. It's an extraordinary idea."[9] (Not extraordinary, utterly unrealistic in my view.) Nevertheless, and regardless of how we define it, the Anthropocene is like nothing that has ever come before!

Continuing human advances, together with our growing numbers, have given rise to one of the most disruptive episodes in the long history of life. Thousands of millions of years were required for the earliest forms of life to elaborate themselves. The evolution of complex animals and a terrestrial flora took place over many millions of years. In bold contrast, and in less than ten thousand years, a bipedal primate has become a species that commandeers over half of the world's terrestrial ecosystems, burns ever more fossil fuel, fills the air with electromagnetic information, and investigates the cosmos by studying the light from distant stars. In fact, in 2014, we blasted flat a high mountain top in Chile to build yet another huge telescope! (The scientific-industrial complex has no intention of curbing its appetite.) Emblematic of our culture, modern societies show no inclination to slow their relentless transformation of the globe. The continuing advance of human technologies and the explosive expansion of our numbers bring us to our final chapter.

Chapter 12

TRILLIONS OF TRANSISTORS: AN UNCERTAIN FUTURE

W e live in extraordinary times! Never before has our planet been so profoundly transformed by a single species—seven billion of us, and counting. Not only have we commandeered more than half of the world's tropical and temperate land areas to feed ourselves and house our growing numbers, but we are disemboweling the Earth itself, building towers and chariots of steel, while feasting on the energy of fossil fuels. In the dead of night, our planet is bejeweled by electric lights from cities large and small. Our technologies are becoming ever-more sophisticated, accelerating communication, enhancing our health, and expanding knowledge of both the world and ourselves. Not unlike the long history of life, we are adding greater complexity to the planet that sustains us.

In fact, we may be witnessing the climax in our planet's long history of getting itself more richly diverse. Though more and more habitats are being converted to human uses, and species are being lost, a huge majority of the world's biological diversity is still with us. And while our own species may be losing some of its languages and smaller cultures, we are inventing ever-more elaborate technologies. Unlike automobiles of a few decades ago, those we purchase nowadays have an automatic transmission, fuel-injected engines, power-assisted steering, cruise control, air bags, and lots more. Thanks to little computers and global satellite positioning, the latest autos direct us to where we want to go. Digitized information, a world-wide Internet, and ever-more sophisticated technologies keep expanding our capabilities.

Surely our species itself is grandly expanding Earth's complexity. If the human mind is among the most complex devices in the natural world, then, logically, adding millions more humans to the planet every month adds to overall complexity, even as our gadget-happy culture produces further innovations every day. By continuously creating and discovering new knowledge, we are expanding the "information content" of our planet. Not unlike complexity, the concept of *information* is difficult to define. Requiring a sender, a receiver, and a shared code for communication, information has been expanding ever since the initiation of life. Even a simple bacterium requires information to absorb its food, control its metabolism, and—most amazing of all—prepare for division into two bacteriums! Likewise, human societies have grown by the creation and interchange of ever-more information. The product of evolutionary dynamics and ecological selection over billions of years, and by expanding our understanding of the world more recently, we have become the most consequential species in the history or life.

HUMAN HEGEMONY

> *We have conquered the biosphere and laid waste to it like no other*
> *species in the history of life. We are unique in what we have wrought.*
> —Edward O. Wilson[1]

Bipedalism freed our arms for many tasks, gesturing hands gave us speech, inter-tribal conflict inflated our brains, agriculture reduced the threat of starvation, and cultural innovations have done the rest. However, it wasn't just us. Historian Paul Conkin argues that our species' ascendancy has benefited from five critical and serendipitous circumstances. First, he cites **climate stability** over the last ten thousand years. Though well documented in Greenland's ice and North Atlantic sediments, we have not witnessed sudden and dramatic climate shifts over the last ten millennia. Second, agriculture has benefited from **rich soils** that were formed over many millions of years. North America's fertile Midwestern Corn Belt is the product of fire-prone prairies having built deep soils over the last twenty million years. Third, our expanding industrial society has been powered largely by **fossil energy** sequestered within the Earth. Oil, coal, and gas are derived from biological remains accumulated and concentrated over more than 500 million years. Without this power base, we could not have become the force we are today. As a fourth condition, Conkin cites the enormous **expansion of human knowledge** over the last three hundred years, a phenomenon initiated by scientific methodology and advanced by the same fossil energy that propels the rest of our society. Finally, Professor Conkin sees the advance of **public health** and medical advances as the fifth driver of our growing numbers and our growing wealth.[2] He might have also mentioned the world's extraordinary biodiversity, allowing us to survive in so many environments and feed ourselves in so many different ways.

HUMAN TECHNOLOGIES ACCELERATE

Eukaryotic cells followed upon two thousand million years of bacterial pioneering. Another thousand million years would pass before animals became larger and more active. Vertebrates ventured onto the land around 370 million years ago (mya), with mammal-like forms appearing around 220 mya. Primates probably arose 60 mya, with apes developing about 20 mya, and our own ancestors becoming upright 6 mya. Humans invented agriculture 10,000 years ago, learned to make steel 3,000 years ago, circumnavigated the globe 500 years ago, improved the steam engine 200 years ago, began flying airplanes 100 years ago, and walked on the Moon 45 years ago.

Not unlike the advance of natural complexity over geological time, human technologies have become ever-more elaborate over these last few thousand years. Today, following both a scientific and an industrial revolution, human technologies are expanding more and more rapidly. "More than anything else technology creates our world. It creates our wealth, our economy, our very way of being," declares W. Brian Arthur.[3] Very much like the natural world over its long history, human technology seems to be elaborating itself in ever shorter time frames.

Technological innovations provided platforms from which further advancement proceeded. Stone tools, woven fibers, and wooden spears may have been the first significant human artifacts. Language—a communication technology—and close social bonding accelerated our advance. Cooking enhanced our caloric intake. A transformational symbiosis, agriculture provided us with more food more reliably. Since then, and by using *intelligent design* within purposeful technological advance, humans have embarked on a trajectory unlike any that has come before. With paper, Islamic scholars gathered, translated, and elaborated knowledge. Later, and again with paper, Gutenberg's printing technology made knowledge more widely accessible. Finally, applying scientific analysis to technological challenges—and with fossil fuels to burn—we humans have become Earth's most successful species ever.

TRILLIONS OF TRANSISTORS

In 1965, Gordon E. Moore published a paper in which he predicted that the number of transistors that could be placed inexpensively on an integrated circuit would approximately double every two years. Amazingly, his prediction held true over many years! Laser photolithography, using sharp ultraviolet light, helped propel this advance in computational power. Beginning with the invention of the integrated circuit in 1958, and thanks to large numbers of tiny transistors, our ability to process information has escalated each and every year. Inevitably, physical and economic constraints halted Moore's Law around 2012. Not to worry: networked processing and many other strategies propel us forward. Today, digital integrated circuits can contain millions of logic gates and many microprocessors. Using an architecture similar to our own neural networks, a "brain chip" with over five billion transistors has recently been developed to analyze visual data. This chip can determine whether a dog, a pedestrian, a bicyclist, or an automobile is crossing an intersection. The digital-computing revolution has been an amazing advance, but there have been costs as well as benefits. Complex computational devices require diverse mineral resources to construct, especially "rare earth" elements. Our many technologies require massive amounts of energy to keep running. Worst of all, many of these advances reduce the need for human labor. Robots do not require vacations or pension plans and—if properly programmed—will never go on strike.

All told, and with the help of trillions of transistors, we have witnessed a major revolution in our ability to process, manipulate, communicate, and retrieve information. Today's desktop computers can do calculations that required a room full of computers in the 1950s. Medicine now has imaging capabilities that can see our insides without cutting us open. Genomic analyses are helping us understand our diseases and discovering thousands of bacterial lineages we never knew before. Elaborate computer models help predict the weather. Within seconds, search engines bring us the information we seek. Internet use rose from about 1 percent of the world's population in 1994 to over 35 percent in 2014. We've gone from 50 billion emails in all of 2006 to about 190 million emails sent and received each day in 2014. Societies are changing as technology becomes ever-more central to our lives. More importantly, nobody really planned this grand progression: it all just happened.

> *The ongoing dilemma of technology, then, will never leave us. It is an ever-elaborate tool that we wield and continually update to improve our world; and it is an ever-ripening super organism of which we are but a part, that is following a direction beyond our own making.*
>
> —Kevin Kelly[4]

In an insightful analysis, Kevin Kelly claims that technology has its own *inner drive* over which we have little control. Like a swift river, ever-expanding technologies carry our society forward into an uncertain future. Kelly believes this trajectory, though causing many problems along the way, will lift our species to new heights, perhaps even some kind of immortality. While his assessment of on-rushing advance is clear enough, I do not share his utopian vision. Kelly fails to mention that ever-accelerating technologies have an ever-greater appetite for energy. He does not discuss the enigma of greater industrial productivity resulting in fewer jobs. He fails to consider the greater susceptibility of more complex and interdependent technologies to sudden system-wide failures. (His notion that "our genes are evolving 100 times faster than in pre-agricultural time" is absurd.[5] Actually, our brains seem to have been getting smaller since the expansion of agriculture.) Though Kelly discusses both the varied and unpredictable negative consequences of technological advance, he worries little about the biosphere that supports this onward rush. Kelly is an optimist, and I am not. I believe that our technologies and our numbers are eroding the world's biodiversity and that, by altering the biosphere itself, threaten our own long-term future. Nevertheless, Kelly's richly argued thesis is convincing: technology has its own unrelenting forward momentum.

AN EVER-INCREASING APPETITE FOR ENERGY

Expanding complexity, whether biological or technological, contradicts the second law of thermodynamics: sooner or later everything runs down. Capturing the energy of sunlight has allowed biology to ignore the second law and continue running uphill over a very long time. Dead, buried, and concentrated over millions of years, fossil fuels began the Industrial Revolution and continue to propel the progressive advancement of our many technologies.

We can no longer live without electric lights, refrigerators, or television. Today, satellites relay TV signals and personal messages all around the globe. Encyclopedias and scientific journals are available on the Internet with only a few key strokes. Looking for information? Google it! (One estimate has ten Google searches using the electricity equivalent to one sixty-watt bulb burning for twenty-eight minutes; it's work, not magic.) Meanwhile, more than four million people are flown around our planet each and every day. It takes around one hundred thousand pounds of fuel to propel a jet liner across the Atlantic, but who's concerned? Our dependence on technology has us burning *around 86 million barrels of oil each and every day*, and that figure is projected to increase each year by an additional million barrels per day. Worldwide consumption of coal doubled between 1980 and 2010. We burned around five billion tons of coal in 2015. These monstrous volumes of oil and coal, added to gas burning, wood burning, cement making, and our increasing numbers, suggest that our hopes for reducing carbon emissions are little more than science fiction.

Worldwide, both energy production and CO_2 emissions increased by about 10 percent between 2004 and 2008. Canada failed to achieve its Kyoto-agreed CO_2 emissions limit largely because of the extraction of oil from its tar sands. Thanks to increasing energy demand in China, and despite slight decreases in Europe and the United States, we humans are emitting about eleven billion tons of CO_2 a year. Of this, around half is absorbed by the oceans and biosphere, the rest is added to our atmosphere. If you think driving an electric car will reduce your carbon footprint, think again. Electric cars charged by older coal-fired generating plants result in about as much CO_2 emission as does a gasoline-fueled car. Meanwhile, *fracking* gives us previously inaccessible oil and gas and keeps the party going—at slightly higher costs.

Reasonable projections estimate that planet Earth may become home to nine billion of our kind by 2050.[6] All the while, our appetite for more comfortable and exciting lives expands. The world's economy grew by as much in the last decade of the twentieth century as the *world's entire economy in 1900!* We may be driving two billion motorized vehicles in 2020. That's great news for personal freedom and commerce, but very bad news for our atmosphere. All this energy consumption has made our lives much more comfortable, a lot more entertaining, and our world ever more complex. Also, as Kevin Kelly argues, we cannot slow the pace of technological progress, or even determine its course. China and India are building many new coal-fired electric plants; their people want electric lights, TV, and refrigerators. Our quest for knowledge also demands more energy, as we attempt to cure our every ailment, design more versatile materials, and prepare to visit Mars.

Back here in the United States, we are using around 38,000,000 disposable plastic bottles each and every day, and billions of plastic bags as well. Near our Midwestern cities, richly productive agricultural lands have given way to an expanding suburbia of "McMansions." These castle-like domiciles are adorned with a three-car garage on the outside and cathedral ceilings on the inside; their ample volume requires more energy to cool in summer and heat in winter. Entrepreneurs may soon be offering short space flights for the wealthy; the carbon footprint of this new diversion is never mentioned. In an authoritative review of our energy future, Nobel laureate Robert Laughlin writes, "the earth's capacity to render up unimaginably large amounts of oil, gas, and coal on demand is a fundamental premise of modern civilization." He believes fossil fuels can power us for another two hundred years, yet ends his text on an optimistic note: "the transition away from fossil fuels is likely to be positive [but] the transition itself could be terrible."[7] Most people seem to agree: if technology created these problems, technology can solve them; everything will turn out just fine.

A VERY OPTIMISTIC SPECIES!

> *Optimism has been central to the process of human evolution. It permitted us to attack wild animals when we hunted and to have some confidence in seeds, soil, water and sun when we planted ... anticipating optimistic outcomes is as much a part of human nature, of the human biology, as the shape of the body.*
>
> —Lionel Tiger[8]

> *The optimism bias stands guard. It is in charge of keeping our minds at ease and our bodies healthy.*
>
> —Tali Sharot[9]

Optimism is surely one of our species' most distinctive characteristics, and clearly a consequence of natural selection. Put simply, optimists were more likely to cope and reproduce than pessimists! Whether as a community or as individuals, optimism has helped us rebound after every calamity, brush ourselves off, and continue on into an uncertain future. **Religion**—a vital part of every society—strengthens us in the face of adversity, commits us to close cooperation, and forms the foundation of our hope. We humans are genetically and culturally programed to be optimistic—and that's a problem.

Unfortunately, **optimism** comes with a close companion: **denial!** "Things really aren't that bad," we tell ourselves. "Global warming is just another climate cycle that we can't do much about. Listen up; it will improve agricultural production in colder climates." In their two fine books on technology, neither W. Brian Arthur nor Kevin Kelly mention the costs—in either energy consumption or environmental services—that continuing industrial expansion demands. These authors and just about everybody else believe that things are only going to get better.

We lurch forward, armed with the same optimism that has made us masters of the planet. Addicted to the comforts of our recent success, we are altering the atmosphere, depleting ocean fisheries, degrading soils, lowering water tables, and consuming ever more energy. Anyone concerned about our future is quickly labeled as having a "doomsday mentality." Successful business people are especially disdainful of negative forecasts. Being successful in business means being bright, being dynamic, being optimistic, and having been lucky. Such leaders are insistent on rushing *development* ever onward.

We do have optimistic technological possibilities. If we could mimic photosynthesis and tear apart the water molecule cheaply, we might live in a hydrogen-powered economy—where the product of combustion is water! Greatly improved batteries might allow us to store electricity more easily—solving the problem of irregularity in wind and solar power. Hydrogen fusion is the process that keeps stars "burning" brightly. Unfortunately, containing the millions of degrees in which hydrogen atoms can fuse has resisted half a century of scientific effort; worse yet, such extreme temperatures may produce radioactive isotopes in the containment structure. In the 1960s, we were assured that hydrogen fusion was only thirty years away; today, hydrogen fusion is still "thirty years away." Genetically modified crops should increase crop productivity, but only if soils are properly maintained. For optimists, there is no end to progressive possibilities.

To be sure, there are strong historical arguments for an optimistic future. Economist Julian Simon declared that everything was getting better and better. Using data beginning in the Middle Ages, Simon and his colleagues showed that health had improved, violence was reduced, ownership of homes and material goods had much expanded, and individual freedoms were now widespread. Indeed, the Industrial Revolution, market capitalism, and modern medicine have given people a far better life than in centuries long past.[10] However, Simon and his associates had nothing to say about declining fish stocks, or the notion that industrial activity might change the weather. They viewed any attempt to restrain the capitalist juggernaut as a diminution of both our freedoms and our future. Julian Simon didn't live long enough to witness China's phenomenal growth or the dramatic changes in the Arctic's climate. From my reading, Simon was a champion of individual rights and economic freedoms; more fundamentally, he was an optimist.

We humans have had to be optimistic, whether dealing with the ordinary travails of life or an uncertain future. Preachers who remind us of misery and suffering offer hope at the end of every sermon. Politicians who warn of a difficult future are soon replaced by those who promise better times. In 2006, Thomas Friedman published a book subtitled *A Brief History of the 21st Century*. The first two hundred pages are a tour de force, making clear how information transfer and digitization has linked and "flattened" the world's economies. Unfortunately, the author all but ignored the natural world. It isn't until page 293 that we hear anything regarding the environment. A glorification of economic globalization, this tract omits many crucial aspects of our new century. Only near the end did the deliriously optimistic author finally write, "At worst we are going to set off a global struggle for natural resources and junk up, heat up, garbage up, smoke up and devour up our little planet faster than at any time in the history of the world."[11] I disagree: it is not "at worst," it is *right now!*

Nowhere in Friedman's *The World Is Flat* do we read anything regarding our ongoing population explosion. I quote Friedman—one of our most highly regarded journalists—as an example of how totally blind contemporary journalism has been to the demographic disaster unfolding around us. To be fair, Friedman addressed many of these issues in his more recent book, *Hot, Flat, and Crowded*, but this text focuses mostly on our energy needs and says little about our escalating numbers. My point is that journalists, economists, and many other "experts" are and have been blind to our planet's greatest dilemma.

THE HUMAN POPULATION EXPLOSION

> *Some scientists, amateur astronomers, and Hollywood filmmakers look fearfully to the skies for civilization-ending bolides. They should look inward. We are the meteor.*
>
> —Eric Roston[12]

In 1997, the *New York Times Magazine* published an article titled "The Population Explosion Is Over!" Data cited by the author were accurate—only he forgot to mention Africa, the Muslim world, most of Asia, and Latin America. Like Simon, and using very specific data sets, the author was arguing for unrestrained personal freedom and economic growth.[13] Since then, and for most journalists, human population growth is yesterday's news—something we need no longer worry about. And why not? Rates of human population growth have declined around the world. True enough, but our numbers continue to expand! This may seem counterintuitive, but three important factors are growing our numbers. First, many more women are living today—with a lower birth rate but producing *almost as many babies* as did fewer women with a higher birth rate during the 1980s. The second factor is the huge number of young people in the pipeline! Young women who are now entering their reproductive years are the second factor propelling population growth forward, in what is called **demographic momentum**. This makes clear why China's population continued to increase for three decades after draconian population-control policies were introduced. The third factor is that the current rate of population growth is still far above replacement level. All three factors explain how our numbers grew from six billion to seven billion between 1999 and 2012.

Explosion is, in fact, the correct word for what's been happening over these last two hundred years. The Indian subcontinent probably numbered 125 million people in 1750; they reached 1,180 million in 2010. Mexico probably had around 13 million people in 1900, and has 112 million today. The population of the Middle East has almost tripled in the last thirty years. High growth rates continue in many poor countries. Recent estimates have the average woman bearing six children in Somalia, Afghanistan, Yemen, and Niger. Even if two of these six children fail to reach adulthood, these populations will double their numbers in twenty-five years! Hunger and malnutrition already threaten these same regions. Sadly, many of the poorest countries of the world continue with high fertility rates.

> *What could be more wicked and heedless of human welfare than to bring more and more children into an overcrowded world, a world so impoverished by overexploitation that they hardly have a fair prospect of a satisfactory life?*
>
> —Alexander Skutch[14]

Here we are, the smartest species ever, racing down the very same path as lesser creatures and earlier societies. Yes, population growth is below replacement level in a few nations, but that's not the big picture. Despite a steady decrease in the rate of population growth, human numbers continue to expand. When I taught in Ethiopia in the early 1960s, that country's population was estimated to be around 22 million; today's estimate is 100 million. Recently, Peter Gill visited a Muslim family in western Ethiopia with fifteen children; such traditions are difficult to change.[15]

Our species' overall annual growth rate peaked at 2.19 percent in 1963 and fell to 1.13 percent in 2015. Annual growth in numbers peaked around 1982 at 88 million additional people, and was well over 70 million in 2015. Despite these simple facts, the UN Millennium Development Goals, designed to decrease world poverty 50 percent by 2015, failed to include population growth. Similarly, two recent books regarding the economics of climate change had nothing to say about our increasing numbers.[16] With ever more young women coming into their reproductive years, and despite decreasing rates of growth, current estimates have our numbers growing by more than 70 million each year for many years to come.

In 1798, Thomas Malthus proclaimed a fearful truth: "Population, when unchecked, increases in geometrical ratio. Subsistence only in an arithmetical ratio." (Arithmetic is 1, 2, 3, 4, 5; geometric is 1, 2, 4, 8, 16.) He warned of "the constant tendency in all animated life to increase beyond the nourishment prepared for it." Accordingly, our planet's agricultural productivity cannot keep pace with human population growth, and famines must follow.[17] Thanks to powerful agricultural machinery, plenty of fossil fuel, industrial fertilizers, and chemical biocides, Malthus' dire prediction has been averted. The recent "Green Revolution" has been able to feed many more people, using better plant varieties, ample water, industrial fertilizers, and chemical biocides. Regions able to access these resources have benefited greatly; because of water limitations, Africa has not.

Contrary to Malthus, rather than wide-spread starvation, our increasing numbers suffer from something more insidious: growing unemployment! Pakistan's population grew by 50 percent in the last thirty years; large numbers of its young people cannot find employment.[18] Economists claim that we need greater productivity to propel prosperity forward. But doesn't *greater* productivity result in *fewer* workers producing as many goods? Digital and robotic systems are reducing the need for human labor. Again, journalists ignore increasing human numbers, even as they report on social strife around the world.

The culture of journalism is basically a political culture that is not particularly hospitable—that is, in fact, institutionally arrogant— toward nonpolitical areas of coverage.

—Ross Gelbspan[19]

Even in a biological journal reviewing four books addressing the global environmental crisis, neither the reviewer nor the books' authors seem concerned about our escalating numbers.[20] Serious famines have been averted, but there is something less draconian to worry about. A massive surge of young people are entering a world of only limited opportunity. Rising unemployment throughout the world portends social instability, ethnic conflict, and relentless emigration. In spite of falling birth rates, 2011 welcomed about 135 million babies into the world, even as 57 million of us departed—a net increase of 78 million.[21] Thanks to 2011 alone, we have *78 million additional mouths to feed, schools to expand, and jobs to create* in the coming decades!

And while the human population explosion is doing just fine, biodiversity is in decline. Two rhinoceros subspecies were declared extinct in 2011: one in West Africa, the other in Vietnam, very likely taken down by hunters. In contrast, the Yangtze River dolphin is the first large mammal species lost to environmental degradation. In 1980, Africa was estimated to have 76,000 lions; this estimate plummeted to 35,000 in 2014. Migratory birds have diminished in great numbers over the last five decades and are continuing to decline.

As a botanist photographing natural scenery around the Western Great Lakes for more than four decades, my observations over the last decade are worrisome. Small areas—less than an acre—within old forest preserves have been subject to severe damage; even as the trees around them are undamaged. Meteorologists call these "microbursts," and they seem to be more common in what I like to call "Global Storming." Dead ash trees (*Fraxinus spp.*) now litter our forests, felled by the emerald ash borer, a beetle recently arrived from China. More troubling is seeing fewer and fewer pollinators in our prairies. Even bright orange monarch butterflies are becoming less common. I suspect that agricultural biocides, however minimal their residues, are beginning to erode our fauna. Our biosphere is suffering from our success.[22]

THE BEST OF TIMES, THE WORST OF TIMES

After a long history of ever-increasing complexity, a single species has made itself the dominant life form on planet Earth. Progressive evolutionary trends, coupled with cultural advance, have finally created a "master generalist." We are that species and our accomplishments are staggering. We have walked on the Moon and explored our solar family with interplanetary probes, making new discoveries millions of miles away. Astronomy has given us insights into the nature and history of the cosmos. Our understanding of our own planet and its life forms has become ever-more revelatory. We are beginning to comprehend the many biochemical networks supporting the living cell, and how interactive neural networks form our minds. Advancing in agriculture, technology, and science, our lives are getting better all the time.

Humans—most of us—have never had it so good. In technologically empowered societies, we are healthier and better fed; we have greater mobility and more ways of entertaining ourselves; we are living longer. Sanitation, cleaner water, better nutrition, vaccinations, antibiotics, and modern medical interventions have extended average life expectancy. Infant mortality is greatly reduced; starvation is countered with shipments of food from afar. Modern agriculture is *fixing* more nitrogen than is being fixed by all the rest of the biosphere, helping feed our growing numbers but changing nitrogen cycling globally. Each day, aircraft carry over four million passengers, and tons of time-sensitive cargo, even as satellites relay electromagnetic information around the globe. Today we find information, communications, and entertainment on the Internet, we watch TV for hours, and we fill huge stadia for sporting or musical events. By using about 30 percent of the planet's primary productivity, and burning prodigious amounts of fossil fuel, we have enriched our lives in ways unimaginable to earlier generations. Stated simply, we are the most *successful* species the world has ever seen. These are the "good times."

The bad news is that many other species are in decline, and the biosphere itself is reeling. One might argue that our species has been increasing its fitness at the expense of most all the others. There are exceptions, of course: wheat, maize, rice, hogs, cattle, and our other favorites are doing well. Twenty billion chickens are helping feed over seven billion humans, even as wild birds decline. World biodiversity, which fostered our success, is being diminished by our success.

With human energy use significantly impacting our atmosphere, our refuse polluting both land and sea, and our numbers continuing to expand, a sixth great extinction in life's long history is underway. The last 500 million years have been witness to five "mass extinctions." This **sixth extinction** cannot be blamed on suffocating volcanism nor an extra-terrestrial jolt. This time a single biological species is the cause of planet-wide decline. Challenging Lovelock's Gaia hypothesis, Peter Ward argues that the same selective forces that drive all species to exploit their environments must inevitably result in environmental degradation.[23] We do not live in a world of happy cooperators, Ward argues; every species has its own selection-driven agenda. Technologically empowered *Homo sapiens* are the biggest thing to hit the planet since an asteroid put the dinosaurs out of business. Crocodiles made it through that last grand disaster; they are not likely to survive the ecological changes we have only just begun.

Human impact can be seen just about everywhere: thousands of square miles of maize, soybean, and wheat cultivation in North America's central plains, carefully sculpted terraces of rice in southeast Asia, pasturelands burning in dry-season Africa, and electrified communities glowing in the dark of night around the globe. In 2012, we harvested around 45 million metric tons of palm oil—on land that once supported lowland tropical evergreen forest. We are behaving like every other species on the planet: we are commandeering more resources to produce ever greater numbers of our own kind, even as we live more lavishly.

Whether red tides in the sea, hordes of migratory locusts, or an outbreak of microbial disease, the patterns are similar. When a species—any species—suddenly finds the resources to multiply, multiply it does. That's the uptick. The downtick comes as the growth spurt runs out of resources. Variously called **boom-and-bust** or **overshoot-and-collapse**, these patterns are familiar in the natural world: exponential population growth followed by resource exhaustion and a population crash. Such cycles often incur nasty environmental effects: poisoned waters in a red tide or devastated vegetation where locust swarms have fed. Human cultures have also expanded in good times or in supportive landscapes, but then, with prolonged drought or as soils become exhausted, food production faltered and societies collapsed. Such scenarios played themselves out in ancient Mesopotamia, in the Mayan lowlands, at Angkor in Cambodia, on Easter Island, and in many smaller communities.[24]

But here we are, the smartest species ever, racing down the very same path as lesser creatures and earlier societies. Yes, population growth is below replacement rate in a few nations, but that's not the big picture. *Human numbers continue to expand, together with an insatiable appetite for more comfort and diversions!*

DIMINISHING ECOSYSTEM CAPITAL

> *The species losses we are now experiencing may foretoken the loss of genera and functional groups, and beyond those, of self-sustaining networks and nutrient cycles. Ecosystems as we know them will be lost, and so too will be nature's services.*
>
> —Simon A. Levin[25]

As has been the case with so many civilizations in the past, our society is fragile. Grain production appears to be reaching a plateau worldwide.[26] World soybean production amounted to 16 million metric tons in 1950, and reached 200 million tons in 2005. Many of these soybeans are making pigs fatter, and sausages-for-breakfast more affordable. Demands for meat-rich diets, electric lighting, and a refrigerator for every family will burn ever more fossil fuel. Over the past few decades, China has rescued more of its people from poverty and expanded its industrial prowess faster than ever before in human history. Powered by fossil fuel, China now leads the world in greenhouse gas emissions. Add these trends to claims by the desperately poor around the world, and we cannot help but continue to expand energy consumption. Also, because of greater interconnectedness, we are increasingly vulnerable to sudden failure; think of electric grids, linked economies, and ocean-traversing oil tankers. The financial meltdown of 2008 was such a collapse. (Deutsche Bank foreclosed on 1,000 homes in Cleveland.) Meanwhile, millions of refugees flee war zones, seek employment, and challenge the stability of more prosperous nations.

Capitalism is founded on an optimistic belief in progress; people who invest their capital will gain profits in a better future. Using the magic elixir of *money*, we can convert the value of food, real estate, labor, or machinery into savings, debt, or just about anything else. Unfortunately, today's globalized capitalism often displaces environmental costs to distant ecosystems.[27] Few of us have witnessed tropical forests being felled to produce oil palms, bananas, plywood, or beef cattle. We are eager to purchase affordable products, and we don't worry about how they are obtained. Nor do we really know—or care to know—where all our garbage ends up. All the while, world fisheries decline, even as agricultural soils lose their fertility. (Rather like mining, farming extracts essential elements from local soils and sends produce to distant urban centers.) "We are devouring our very life-support systems, and finding excuses along the way not to care," writes Jeffrey Sachs in a comprehensive analysis of our current predicament.[28] Despite so many negative trends, Sachs remains optimistic, arguing that we can develop sustainable solutions. Vaclav Smil, in a detailed analysis of current threats, is also quite hopeful.[29] Brahma Chellaney, on the other hand, claims that we are facing shortages in the single resource for which there is absolutely no substitute: clean fresh water. While the human population increased by a factor of 3.6 over the twentieth century, our demand for fresh water increased nine-fold. Advancing societies have an appetite for beef, but a pound of beef requires ten times the water that's needed for a pound of wheat.[30] Ocean waters can be made drinkable with desalinization plants, but that requires yet more energy consumption. Nevertheless, and insulated by our innate optimism, few people are fearful of a world of seriously limited resources or damaged ecosystems.[31]

In part, I agree. Yes, there's a major extinction event underway and environments are changing, but we humans will not be among the departed! Our amazingly versatile species survived all the recent glacial cycles, and we did that with only limited technology. Together with rats and roaches, humans will surely persist. Fears of human self-extinction by nuclear war were grossly exaggerated in my opinion; folks in the Southern Hemisphere would be little affected. Likewise, rebellious robots or nasty nano-beasties are science fiction: where might they acquire the energy for making trouble? Today's threat is much less dramatic and much more serious. Billions of people dependent on modern industrial societies are not a good formulation for long-term sustainability. I suspect that if our energy-devouring civilization falters, our species will muddle on—although with reduced numbers and much diminished lifestyles, to be sure. Humans have survived over thousands of generations in the rocky thorn bush of eastern Africa, hundreds of generations along the edge of the Arctic Ocean, and many generations in the back alleys of Calcutta; we are a tough crowd and are likely to be around for a long time to come.

Unfortunately, resource-dependent, energy-demanding societies—focused primarily on wealth acquisition—are unlikely to remain stable over the longer haul. Our unique industrial civilization, like so many that have come before, will probably end in disarray. But this is a very pessimistic view, rarely seen in print or voiced in conversation. Our brains have been *programmed* to imagine a better future; very few people dare to look forward realistically.

Optimistic experts, in contrast, believe that new technologies will solve the problems that modern lifestyles create. This is a quite reasonable assertion, considering the unimagined advances we have made over the last two centuries. Nevertheless, and whichever view you prefer, it does appear that our looming difficulties are the direct consequence of our own best intentions.

GOOD INTENTIONS: UNINTENDED CONSEQUENCES

I start from the bedrock principle that we as a global society need more and more growth, because without growth there is no human development and those in poverty will never escape it.
—Thomas Friedman[32]

The planet on which our civilization evolved no longer exists. The stability that produced that civilization has vanished, epic changes have begun . . . the transition from a system that demands growth to one that can live without it will be wrenching.
—Bill McKibben[33]

Modern societies are inspired by economic systems that demand ceaseless growth: all dedicated to improving our lives.[34] Unfortunately, and as the old aphorism tells us, *"The road to Hell is paved with good intentions."* Few people consider that our most cherished values may be undermining our future. Since earliest agricultural times, we have claimed the *human right* to cut the forest, clear the land, and harvest whatever resources we demanded. Modern medicine has multiplied our numbers, even as everyone seeks to share in the comforts of modern technology. Both by transforming natural ecosystems and burning fossil fuels we are altering the capacity of the biosphere to sustain us. "We are emptying our coal mines into the sky!" warned chemist Svante Arrhenius in 1900. Belching billions of tons of carbon dioxide into the air, our atmosphere is becoming not just warmer but a lot more volatile!

Global warming has many insidious consequences. High mountain glaciers are melting. Capturing fresh water as snow during the wet season, melting glaciers support irrigation systems for billions of people at lower elevations. Worst of all, global warming may produce greater storming in some areas and more severe drought in others. Recently, warmer waters in the northeast Pacific diverted the jet stream, bringing the "Siberian Express" southward into the central and eastern United States, making the winters of 2013–14 and 2014–2015 among the coldest on record. The winter of 2015–2016 is altogether different, thanks to a strong El Niño in the southern Pacific. All the while industry-funded "think tanks" question human-induced climate change with their own small army of *subsidized experts*. Journalists use this "expertise" to counter scientific consensus in a "fair and balanced" manner (cf. Fox News).

Ultimately, love and mercy may prove to be far more ruinous than atomic war or deadly plague. Love produces more babies; mercy nourishes them and helps them grow. Bringing forth children and raising them well are central to human survival and give our life its most rewarding purpose. Curbing our birthrate contradicts both religious teaching and biological imperative. For both the liberal and the conservative, *birth control* is an unacceptable constraint of a fundamental "human right." Yes, most nations are experiencing reduced population growth, but our numbers are still expanding mightily. Sub-Saharan Africa, with many traditions fostering large families, is growing by about 80,000 people each and every day. Anyone who thinks we can reduce our carbon dioxide output or limit our use of non-renewable resources in a world of growing human numbers has lost touch with reality.

In an illuminating and level-headed review of our predicament, Laurence Smith focuses on the northernmost regions of our planet. Smith's book, *The World in 2050*, focuses on four major forces driving our future: (1) **Demographic Growth**, (2) **Increasing Resource Demands**, (3) **Economic Globalization**, and (4) **Climate Change**. He makes clear that the reduction of summer sea ice is the primary driver of rapid climate change around the Arctic Ocean. Ending his book, he asks: "What kind of world do we want?"[35] Seems to me, the dynamics of his four forces will determine our future and not *what we want*! I agree with Elizabeth Kolbert, who concluded her book on climate change with the warning, "It may seem impossible to imagine that a technologically advanced society could choose, in essence, to destroy itself, but that is what we are now in the process of doing."[36]

AN UNCERTAIN FUTURE

> *The primordial blessing "increase and multiply" has suddenly*
> *become a hemorrhage of terror. We are numbered in the billions and*
> *massed together, marshalled, marched here and there, taxed, drilled,*
> *armed, worked to the point of insensibility, dazed by information,*
> *drugged by entertainment, surfeited with everything.*
> —Thomas Merton[37]

Finally, after more than three billion years, Mother Nature's tendency to get more complicated has fashioned a species able to expand its food and energy sources, defeat its pathogens, elaborate its technologies, and multiply its numbers, all without comprehending the long-term consequences. We differ little from other species, all striving to expand their numbers. Humans are operating much as did plagues of desert locusts in biblical times. Perhaps a *master species* is the inevitable culmination of nature's innate progressive tendencies. Propelled by the same subtle forces that have driven biological elaboration over these last three billion years, our species has grandly amplified that process with progressive technological advance. Traditional subsistence farming, supporting millions of families, is being deliberately replaced by large scale industrial agriculture, sending displaced farm families to ever-more congested cities. Both modern technologies and population expansion have made human labor "the world's most overabundant resource"![38] Worldwide unemployment, propagated by both unimpeded free-market globalization and advancing technologies, has left many without the financial security to begin a family and look forward to a fulfilling life. In the face of such hopelessness, religious extremism offers both purpose and immortality, fueling terrorism around the globe. Meanwhile, our optimism insists that modern economies can become sustainable, while contradictory evidence is dismissed as "catastrophic thinking."

In an earlier volume, I suggested that we humans might possess the only radio telescopes currently functioning in our corner of the Milky Way Galaxy. That's a stretch: there are billions of planets out there. However, a huge number of *lucky breaks* have provided us with a stable star, a supportive planet, nutritious food, big brains, and ever-advancing culture. Without a large stabilizing Moon, plate tectonic support for "floating" land masses, expansion of the flowering plants, extinction of the dinosaurs, and grasslands supporting tasty herbivores, we humans simply would not be here. My final argument for the lack of informative radio signals in our galactic neighborhood was that exploitive technological civilizations are unlikely to be sustainable, and they vanish quickly.[39] One can imagine other civilizations having arisen, circling other stars. Perhaps they grew, prospered, invented radio communication, but then made a mess of their biosphere and crashed; surviving only as fragmented bands eking out a living on their plundered planet. The Search for Extraterrestrial Intelligence with radio telescopes (SETI) has detected not a single coherent signal on the interstellar airwaves in over forty years. Might our own radio telescopes shut down when ten billion humans clamor for their fundamental needs?

Our species is transforming our world, but are we capable of managing these changes? Sociologists Philip Smith and Nicolas Howe, using Aristotelian principles, examine *Climate Change as Social Drama*. They point out that "The problem is complex in its causality, widespread in its impacts, and not easily converted into compelling cultural forms that transmit danger and urgency to ordinary people." They hope for a truly compelling social drama "that will change history for *us*."[40] These authors have little to say about our innate optimism or tendency to deny negative human agency. Like so many other *experts*, they discuss the human population explosion *not at all!* The authors do recognize that climate change challenges deeply held notions of *progress and human rights* and is unique in demanding reductions in goods and services, all for the greater good of future generations. But how likely is that?

> *Unless we can ensure that the Economy is kept subservient to our Ecology we will self-destruct.*
>
> —Roger Short[41]

Traditional economics preaches perpetual growth, even though nothing, save God or the universe can possibly be perpetual—and there's some doubt about the universe.

—Alan Weisman[42]

The future of humankind is unknowable. We live in a complex universe where many things "can go wrong." Nevertheless, and despite a lot of bad news, most folks have an optimistic vision for our future. Huge improvements in our lives and lifestyles over the last three centuries support such a worldview. In contrast, students of the environment are deeply fearful that our relentless demand for resources will lead to a severely diminished biosphere. Despite such differing viewpoints, I believe we can all agree that recent human achievements are the culminating epiphany in a long history that's been getting ever more complicated. We humans have become—quite literally—the pinnacle of organic evolution. Whether measured by our understanding of the cosmos, the luxury of our lifestyles, or our impact on the biosphere, humans are the most transformative species in the history of life! We have added significantly to the complexity of our planet.

With billions of beetles enlivening our terrestrial ecosystems, and trillions of transistors enabling our technologies, planet Earth supports more complexity than all the other members of our solar system combined. Lots of water, ample land surfaces, escalating biodiversity, and our own splendid ingenuity, have driven this scenario of ever-escalating complexity. Considering all the "lucky breaks" throughout the planet's history, how likely is it to find another heavenly body similarly adorned? I suspect that our galaxy has nothing to equal our globe's complexity within the nearest thousand light years. We are truly very special.

That said, human activities are threatening the very richness and stability of our biosphere. Regrettably, I am a pessimist who sees modern industrial society as unsustainable and "human nature" incapable of diminishing its biological and cultural appetites. Also, I'm worried that, since people cause trouble, more people will inevitably cause more trouble. Ever growing numbers make our future less likely to become better and better. Yes, the Industrial Revolution and its ongoing advancement have given us near-miraculous technologies. Our lives and our planet have been transformed! However, there is no guarantee that similarly inventive miracles will solve future challenges. Even if we could wring hydrogen from water cheaply or achieve the controlled fusion of hydrogen into helium, our resource devouring and ever-growing numbers are unlikely to be sustainable.

Today, a significant number of Christians believe that we are entering the *end times of biblical prophecy*—and that God's promises will be soon fulfilled. Many Muslims believe that increasing chaos will bring the *Return of the Mahdi*—and all will be well. Propelled by our innate optimism, most everybody anticipates a better future.

However, and regardless of how one views our future, I believe there is one further point on which hopeful optimists, religious believers, and fearful pessimists can all agree. Rephrasing the Ancient Chinese Curse:

We live in interesting times!

NOTES

INTRODUCTION

1. Diversity in biological systems can be best defined by the number of elements within each system. In this sense, a genus or a habitat with many species is more diverse than a genus or habitat with few.

2. See A. Begossi et al., "Are Biological Species and Higher-Ranking Categories Real? Fish Folk Taxonomy on Brazil's Atlantic Forest Coast and in the Amazon," *Current Anthropology* 49, no. 2 (April 2008): 291–306.

3. For a broader perspective, see Eric W. Holman, "How Comparable Are Categories in Different Phyla?" *Taxon* 56, no. 1 (February 2007): 179–84. For a recent estimate on species numbers, see Mark J. Costello, Robert M. May, and Nigel E. Stork, "Can We Name Earth's Species before They Go Extinct?" *Science* 339, no. 6118 (January 25, 2013): 413–16.

4. A multi-institutional effort is currently underway to make information available regarding all 1.8 million described species. To access this effort see the *Encyclopedia of Life* website.

5. Paul Oliver et al., "Cryptic Diversity in Vertebrates: Molecular Data Double Estimates of Species Diversity in a Radiation of Australian Lizards (*Diplodactylus*, Gekkota)," *Proceedings of the Royal Society B* 276, no. 1664 (June 7, 2009): 2001–2007; David R. Vieites et al., "Vast Underestimation of Madagascar's Biodiversity Evidenced by an Integrative Amphibian Inventory," *Proceedings of the National Academy of Sciences of the United States of America* 106, no. 20 (May 19, 2009): 8267–72.

6. A recent study in China gives strong support to the notion that the species richness of land vertebrates and vascular plants are concordant; see Haigen Xu et al., "Biodiversity Congruence and Conservation Strategies: A National Test," *BioScience* 58, no. 7 (July/August 2008): 632–39.

7. For an entertaining and strongly opinionated review, see Richard Greenberg, *Unmasking Europa: The Search for Life on Jupiter's Ocean Moon* (New York: Copernicus, 2008). This delirious author even illustrates animals creeping around the fissures of Europa's icy crust (p. 249).

8. For a nice overview regarding the physical and chemical nature of our universe, see John Gribbin, *The Origins of the Future: Ten Questions for the Next Ten Years* (New Haven, CT: Yale University Press, 2006).

9. For a short review of the Moon's origin, see Alex N. Halliday, "Planetary Science: Isotopic Lunacy," *Nature* 450, no. 7168 (November 15, 2007): 356–57; and for more recent theories see Alex N. Halliday, "The Origin of the Moon," *Science* 338, no 6110 (November 23, 2012): 1040–41.

10. Kevin J. Walsh, "Asteroids: When Planets Migrate," *Nature* 457, no. 7233 (February 26, 2009): 1091–93.

11. Richard Kerr, "Planetary Two-Step Reshaped Solar System, Saved Earth?" *Science* 332, no. 6035 (June 10, 2011): 1255.

12. For a detailed technical book regarding life-supporting planets, see James Kasting, *How to Find a Habitable Planet* (Princeton, NJ: Princeton University Press, 2010).

13. John Chambers and Jacqueline Mitton, *From Dust to Life: The Origin and Evolution of Our Solar System* (Princeton, NJ: Princeton University Press, 2014), p. xv. A fine review of the Sun and its family.

14. In fact, changes in the Earth's degree of tilt, changing eccentricity of its orbit, and changes in the time of year when we are closest to the Sun, have contributed to recent cycles of glaciation. For an overview, see Mark A. Maslin and Beth Christensen, "Tectonics, Orbital Forcing, Global Climate Change, and Human Evolution in Africa: Introduction to the African Paleoclimate Special Volume," *Journal of Human Evolution* 53, no. 5 (November 2007): 443–64.

15. Eric Roston sees plate tectonics as central to sustaining life on planet Earth, second only to sunshine. See *The Carbon Age: How Life's Core Element Has Become Civilization's Greatest Threat* (New York: Walker, 2008).

16. Venus doesn't have this particular problem because of its dynamic atmosphere. Without its dense heat-trapping atmosphere, Venus would have one side broiling under the hot Sun over many months, as the other side resides in deep frigidity. With or without its suffocating atmosphere, Venus could not support a biosphere.

17. I have even argued that, thanks to all the lucky breaks in our long history, we may have the only radio telescopes currently functioning in our galaxy; see *Perfect Planet, Clever Species: How Unique Are We?* (Amherst, NY: Prometheus Books, 2003).

18. Though I see "astrobiology" as a clever ploy, helping the Astronomical-

Industrial -Complex continue sucking dollars out of our national budget, Lucas John Mix has written a fine book, reviewing the physical requirements for life: *Life in Space: Astrobiology for Everyone* (Cambridge, MA: Harvard University Press, 2009).

19. For a readable narrative regarding our universe and ourselves, see Christopher Potter, *You Are Here: A Portable History of the Universe* (New York: HarperCollins, 2009).

20. For a fine rejoinder to arguments of those supporting "intelligent design," and lots of evidence supporting the theory of our having evolved over time, see Jerry Coyne, *Why Evolution Is True* (New York: Viking, 2009).

21. Nicholas Wade, "Evolution All Around," *New York Times Book Review*, October 11, 2009: 22.

22. Michael J. Benton, "The Origins of Modern Biodiversity on Land," *Philosophical Transactions of the Royal Society B* 365, no. 1558 (November 27, 2010): 3667–79; John C. Briggs, "Species Diversity: Land and Sea Compared," *Systematic Biology* 43, no. 1 (March 1994): 130–35.

23. Christopher B. Field et al., "Primary Production of the Biosphere: Integrating Terrestrial and Oceanic Components," *Science* 281, no. 5374 (July 10, 1998): 237–40.

24. For a technical but well-written overview of photosynthesis, see Oliver Morton, *Eating the Sun: How Plants Power the Planet* (New York: HarperCollins, 2007).

25. Brian Groombridge and Martin D. Jenkins, *World Atlas of Biodiversity: Earth's Living Resources in the 21st Century* (Berkeley, CA: University of California Press, 2002), p. 11.

26. Vaclav Smil, *The Earth's Biosphere: Evolution, Dynamics, and Change* (Cambridge MA: MIT Press, 2002), p. 7. It has also been suggested that 90 percent of marine biomass is microbial; see Dennis Normile, "Counting the Ocean's Creatures, Great and Small," *Science* 330, no. 6000 (October 1, 2010): 25.

27. See the short review and included references by Amber Dance, "What Lies Beneath?" *Nature* 455, no. 7214 (October 9, 2008): 724–25.

28. Rob R. Dunn reviews our efforts at understanding the diversity of life in his book *Every Living Thing: Man's Obsessive Quest to Catalog Life, from Nanobacteria to New Monkeys* (New York: HarperCollins, 2009). Championing lone voices that spurred recent breakthroughs, together with personal field experiences, help animate this very readable survey. Unfortunately, devoting

many pages to a hypothetical "nanobacteria" and to an imaginative astrobiology seems more science fiction than science fact.

29. Estimating the total number of species continues to be controversial, and many estimates seem wildly overstated. For a short review, see Robert M. May, "Tropical Arthropod Species, More or Less?" *Science* 329, no. 5987 (July 2, 2010): 41–42.

CHAPTER 1: BILLIONS OF BEETLES

1. Hanno Sandvik, "An Inordinate Fondness for Mecopteriforma," *Systematics and Biodiversity* 4, no. 4 (December 2006): 381–84.

2. The "type specimen" is important in plant and animal nomenclature. This is the specimen to which the new name is forever linked. Populations vary: similar species may grow together; technical descriptions can be subject to different interpretations. Thus, when questions arise as to how to apply a specific name precisely, the type specimen serves as a touchstone.

3. For a short general discussion, see John Tyler Bonner, "Matters of Size," *Natural History* 115, no. 9 (November 2006): 54–58. For a more technical review, see Ethan P. White et al., "Relationships between Body Size and Abundance in Ecology," *Trends in Ecology and Evolution* 22, no. 6 (June 2007): 324–30.

4. Gilbert Waldbauer provides a nice overview in *Millions of Monarchs, Bunches of Beetles: How Bugs Find Strength in Numbers* (Cambridge, MA: Harvard University Press, 2000).

5. Toby Hunt et al., "A Comprehensive Phylogeny of Beetles Reveals the Evolutionary Origins of a Superradiation," *Science* 318, no. 5858 (December 21, 2007): 1913–16.

6. Michael Kaspari and Bradley Stevenson, "Evolutionary Ecology, Antibiosis, and All that Rot," *Proceedings of the National Academy of Sciences of the United States of America* 105, no. 49 (December 9, 2008): 19027–28.

7. The numbers of little wasp species appear to be grossly underestimated; see, for example, M. Alex Smith, et al., "Extreme Diversity of Tropical Parasitoid Wasps Exposed to Iterative Integration of Natural History, DNA Barcoding, Morphology, and Collections," *Proceedings of the National Academy of Sciences of the United States of America* 105, no. 34 (August 26, 2008): 12359–64.

8. Peter J. Mayhew, "Why Are There So Many Insect Species? Perspectives from Fossils and Phylogenies," *Biological Reviews* 82, no. 3 (August 2007): 425–54.

9. Erwin went on to estimate huge numbers of undiscovered species around the world, but these projections are probably overblown. For a short review of Erwin's work, see Rob Dunn, *Every Living Thing: Man's Obsessive Quest to Catalog Life, from Nanobacteria to New Monkeys* (New York: HarperCollins, 2009), pp.77–82. His nanobacteria, however, may be more fiction than fact.

10. Ryoko Okajima, "The Controlling Factors Limiting Maximum Body Size in Insects," *Lethaia* 41, no. 4 (December 2008): 423–30.

11. David Beerling, *The Emerald Planet: How Plants Changed Earth's History* (New York: Oxford University Press, 2007).

12. Sean B. Carroll, *Endless Forms Most Beautiful: The New Science of Evo Devo* (New York: W. W. Norton, 2005).

13. Jan Schipper et al., "The Status of the World's Land and Marine Mammals: Diversity, Threat, and Knowledge," *Science* 322, no. 5899 (October 10, 2008): 225–30.

14. The ongoing discovery of many new species make 300,000 species a more likely "final tally" for flowering plants. In fact, most new species are being found in "biodiversity hotspots," which we'll be discussing in chapters four and five.

15. Plant family and generic species numbers quoted in this book come from the very useful *Mabberley's Plant Book* (Cambridge, UK: Cambridge University Press, 2008).

16. Brian D. Farrell, "'Inordinate Fondness' Explained: Why Are There So Many Beetles?" *Science* 281, no. 5376 (July 24, 1998): 555–59.

17. I discussed the significance of Angiosperms both for the planet and human history in *Flowers: How They Changed the World* (Amherst, NY: Prometheus Books, 2006).

CHAPTER 2: BACTERIA, EUKARYOTIC CELLS, AND SEX

1. For a recent review of bacteria and their role in modern biological science, see Carl Zimmer, *Microcosm: E. coli and the New Science of Life* (New York: Pantheon Books, 2008).

2. For a terrific but dense review of photosynthesis and its effect on our planet, see Oliver Morton, *Eating the Sun: How Plants Power the Planet* (New York: HarperCollins Publishers, 2008).

3. If the average bacterium is about one micron wide and three microns long, this adds up to a "footprint" of three square microns. With the average printed period having a diameter between three hundred and four hundred microns, as many as 20,000 bacteria should be able to crowd together within the period at the end of this sentence.

4. Michael Gleich et al., *Life Counts: Cataloguing Life on Earth* (New York: Atlantic Monthly Press, 2002), p. 29.

5. For a short overview and references, see David S. Schneider and Moria C. Chambers, "Rogue Insect Immunity," *Science* 322, no. 5905 (November 21, 2008): 1199–200.

6. Ernst Mayr, *Animal Species and Evolution* (Cambridge, MA: Belknap Press, 1963), p. 19.

7. Elizabeth Pennisi, "Researchers Trade Insights about Gene Swapping," *Science* 305, no. 5682 (July 16, 2004): 334–35.

8. James O. McInerney and Davide Pisani, "Paradigm for Life," *Science* 318, no. 5855 (November 30, 2007): 1390–91.

9. For more detailed discussions of species identification among bacteria, see Christophe Fraser, William P. Hanage, and Brian G. Spratt, "Recombination and the Nature of Bacterial Speciation," *Science* 315, no. 5811 (January 26, 2007): 476–80; John Bohannon, "Confusing Kinships," *Science* 320, no. 5879 (May 23, 2008): 1031–33; and David Emerson et al., "Identifying and Characterizing Bacteria in an Era of Genomics and Proteomics," *Bioscience* 58, no. 10 (November 2008): 925–36.

10. For a fine review of the Archaea, see Tim Friend *The Third Domain: The Untold Story of Archaea and the Future of Biotechnology* (Washington, DC: Joseph Henry Press, 2007).

11. Thomas Cavalier-Smith, "Cell Evolution and Earth History: Stasis and Revolution," *Philosophical Transactions of the Royal Society B* 361, no. 1470 (June 29, 2006): 969–1066. See also: Karou Fukami-Kobayashi et al., "A Tree of Life Based on Protein Domain Organization," *Molecular Biology and Evolution* 24, no. 5 (May 2007): 1181–89.

12. Thomas Cavalier-Smith, "Origin of Mitochondria by Intracellular Enslavement of a Photosynthetic Purple Bacterium," *Proceedings of the Royal Society B: Biological Sciences* 273, no. 1596 (August 7, 2006): 1943–52.

13. For a short overview, see Carl Zimmer, "On the Origin of Eukaryotes," *Science* 325, no. 5941 (August 7, 2009): 666–68. For a more technical argument, see

T. Vellai et al., "A New Aspect to the Origin and Evolution of Eukaryotes," *Journal of Molecular Evolution* 46, no. 5 (May 1998): 499–507.

14. For a detailed argument regarding the importance of mitochondria in the history of life, see Nick Lane's *Power, Sex, Suicide: Mitochondria and the Meaning of Life* (New York: Oxford University Press, 2005).

15. Actually, oak trees don't have egg cells. The female oak flower has an ovule with a complex "egg apparatus" containing a number of nuclei. A single nucleus from the pollen tube will unite with an egg nucleus to form the basis of a new oak tree; this is similar to a sperm nucleus uniting with an egg nucleus.

16. For a technical review of current thinking regarding the origin of eukaryotes, see Yonas I. Tekle, Laura Wegener Parfrey, and Laura A. Katz, "Molecular Data Are Transforming Hypotheses on the Origin and Diversification of Eukaryotes," *Bioscience* 59, no. 6 (June 2009): 471–81.

17. For a short review of the evolution of sex, see Carl Zimmer, "On the Origin of Sexual Reproduction," *Science* 324, no. 5932 (June 5, 2009): 1254–56. A more recent summary is provided by Michael Brockhurst, "Sex, Death, and the Red Queen," *Science* 333, no. 6039 (July 8, 2011): 166–68.

18. This mathematical analysis used Linnaean ranks to estimate eukaryotic species-numbers; see Daniel Strain, "8.7 million: A New Estimate for All the Complex Species on Earth," *Science* 333, no. 6046 (August 26, 2011): 1083.

CHAPTER 3: WHAT DRIVES THE FORMATION OF NEW SPECIES?

1. For a discussion of the "design problem" in evolutionary thought, see Michael Ruse, *Darwin and Design: Does Evolution Have a Purpose?* (Cambridge, MA: Harvard University Press, 2003).

2. For a short summary of Darwin's achievement and suggestions for further progress, see Rasmus Grønfeldt Winther, "Systemic Darwinism," *Proceedings of the National Academy of Sciences of the United States of America* 105, no. 33 (August 19, 2008): 11833–38.

3. Ernst Mayr, *Animal Species and Evolution* (Cambridge, MA: Belknap Press, Harvard, 1963), p. 19.

4. The complexity of population dynamics among finches in the Galapagos Islands is described by the two scientists that have been following

them most closely; see Peter R. Grant and B. Rosemary Grant's *How and Why Species Multiply: The Radiation of Darwin's Finches* (Princeton, NJ: Princeton University Press, 2007).

5. Ian Tattersall, "Madagascar's Lemurs: Cryptic Diversity or Taxonomic Inflation?" *Evolutionary Anthropology* 16, no. 1 (January/February 2007): 12–23; C. Schwitzer et al., "Averting Lemur Extinctions amid Madagascar's Political Crisis," *Science* 343, no. 6173 (February 21, 2014): 842–44.

6. B. D. Dow and Mary V. Ashley, "High Levels of Gene Flow in Bur Oak Revealed by Paternity Analysis Using Microsatellites," *Journal of Heredity* 89, no. 1 (January 1998): 62–70.

7. For example, a recent study of fish in Brazil, both on the Atlantic shore and in the Amazon, found that local fishers' folk taxonomy was very similar to that used by biologists. See: A. Begossi et al., "Are Biological Species and Higher-Ranking Categories Real?: Fish Folk Taxonomy on Brazil's Atlantic Forest Coast and in the Amazon," *Current Anthropology* 49, no. 2 (April 2008): 291–306.

8. Rolla Tryon, "Development and Evolution of Fern Floras of Oceanic Islands," *Biotropica* 2, no. 2 (November 1970): 76–84.

9. Jurgen Haffer, "Speciation of Amazonian Forest Birds," *Science* 165, no. 3889 (July 11, 1969): 131–37.

10. Paul Colinvaux and P. E. De Oliveira, "Amazon Plant Diversity and Climate through the Cenozoic," *Paleogeography, Paleoclimatology, Paleoecology* 166, no 1–2 (February 1, 2001): 51–63.

11. Menno Schilthuizen, *Frogs, Flies, and Dandelions: Speciation—The Evolution of New Species* (Oxford, UK: Oxford University Press, 2001).

12. "The Wallace Effect" is discussed in Verne Grant's *Plant Speciation* (New York, NY: Columbia University Press, 1971), pp. 76–84.

13. Actually, a recent study suggests that human lice may not represent two distinct species. See: Jessica E. Light, Melissa A. Toups, and David L. Reed, "What's in a Name: The Taxonomic Status of Human Head and Body Lice," *Molecular Phylogenetics and Evolution* 47, no. 3 (June 2008): 1203–16.

14. William Burger, "Montane Species-Limits in Costa Rica and Evidence for Local Speciation on Altitudinal Gradients," in S. P. Churchill, H. Balsev, E. Ferero, and J. Luteyn, eds., *Biodiversity and Conservation of Neotropical Montane Forests* (Bronx, NY: New York Botanical Garden, 1995), pp. 127–33.

15. Christopher Schneider et al., "A Test of Alternative Models of Diversification in Tropical Rainforests: Ecological Gradients vs. Rainforest

Refugia," *Proceedings of the National Academy of Sciences of the United States of America* 96, no. 24 (November 23, 1999): 13869–73.

16. R. C. Albertson et al., "Phylogeny of a Rapidly Evolving Clade: The Cichlid Fishes of Lake Malawi, East Africa," *Proceedings of the National Academy of Sciences of the United States of America* 96, no. 9 (April 27, 1999): 5107–10.

17. M. Emilia Santos and Walter Salzburger, "How Cichlids Diversify," *Science* 338, no. 6107 (November 2, 2012): 619–20.

18. Tom Tregenza and Roger K. Butlin, "Speciation without Isolation," *Nature* 400, no. 6742 (July 22, 1999): 311–12; Kerstin Johannesson, "Parallel Speciation: A Key to Sympatric Divergence," *Trends in Ecology and Evolution* 16, no. 3 (March 2001): 148–53.

19. D. Luke Mahler et al., "Exceptional Convergence on the Macroevolutionary Landscape in Island Lizard Radiations," *Science* 341, no. 6143 (July 19, 2013): 292–95.

20. Stefanie De Bodt, Steven Maere, and Yves Van de Peer, "Genome Duplication and the Origin of Angiosperms," *Trends in Ecology and Evolution* 20, no. 11 (November 2005): 591–97.

21. For a recent discussion, see the symposium issue beginning with Richard J. Abbott, Michael G. Ritchie, and Peter M. Hollingsworth, "Introduction. Speciation in Plants and Animals: Pattern and Process," *Philosophical Transactions of the Royal Society B* 363, no. 1506 (September 27, 2008): 2965–69, and following.

CHAPTER 4: THE GEOGRAPHY OF SPECIES RICHNESS

1. For an excellent text covering plant and animal distribution around the world, see C. Barry Cox and Peter D. Moore, *Biogeography: An Ecological and Evolutionary Approach* (Malden, MA: Blackwell Publishing, 2005). Using phylogenetic relationships, Ben Holt et. al. propose "An Update of Wallace's Zoogeographic Regions of the World," *Science* 339, no. 6115 (January 4, 2013): 74–77.

2. F. I. Woodward, M. R. Lomas, and C. K. Kelly, "Global Climate and the Distribution of Plant Biomes," *Philosophical Transactions of the Royal Society B* 359, no. 1450 (October 29, 2004): 1465–76. For a comprehensive text, see Susan L. Woodward, *Biomes of Earth: Terrestrial, Aquatic, and Human-Dominated* (Westport, CT: Greenwood Press, 2003).

3. David M. Olson et al., "Terrestrial Ecoregions of the World: A New Map of Life on Earth," *BioScience* 51, no. 11 (November 2001): 933–38.

4. Robin Abell et al., "Freshwater Ecoregions of the World: A New Map of Biogeographic Units for Freshwater Biodiversity Conservation," *BioScience* 58, no. 5 (May 2008): 403–14.

5. Thomas E. Lovejoy, "Biodiversity: What Is It?" in *Biodiversity II: Understanding and Protecting our Biological Resources*, ed. Marjorie L. Reaka-Kudla, Don E. Wilson, and Edward O. Wilson (Washington, DC: Joseph Henry Press, 1997), pp. 7–14.

6. Vojtech Novotny et al., "Why Are There So Many Species of Herbivorous Insects in Tropical Rainforests?" *Science* 313, no. 5790 (August 25, 2006): 1115–18.

7. Matthew Symonds and Christopher Johnson, "Species Richness and Evenness in Australian Birds," *American Naturalist* 171, no. 4 (April 2008): 480–90.

8. Northern boreal forests are a geologically recent formation and may be severely affected by global warming. See: Ralph E. Taggart and Aureal T. Cross, "Global Greenhouse to Icehouse and Back Again: The Origin and Future of the Boreal Forest Biome," *Global and Planetary Change* 65, no. 3–4 (February 2009): 115–21.

9. James M. Dyer, "Revisiting the Deciduous Forests of Eastern North America," *BioScience* 56, no. 4 (April 2006): 341–52.

10. Thomas J. R. Finnie et al., "Floristic Elements in European Vascular Plants: An Analysis Based on *Atlas Florae Europaeae*," *Journal of Biogeography* 34, no. 11 (November 2007): 1848–72.

11. Russell A. Mittermeier, Norman Myers, and Cristina G. Mittermeier, *Hotspots: Earth's Biologically Richest and Most Endangered Terrestrial Ecoregions* (Mexico City: CEMEX, 1999), p. 37.

12. Y. Yang et al., "Biodiversity and Biodiversity Conservation in Yunnan, China," *Biodiversity and Conservation* 13, no. 4 (April 2004): 813–26.

13. For an illustrated overview of Mediterranean vegetation, see: Peter R. Dallman, *Plant Life in the World's Mediterranean Climates* (Berkeley, CA" University of California Press, 1998).

14. F. Miranda and E. Hernandez, "Los tipos de vegetación de Mexico y su classifación," *Boletin Sociedad Botanica de Mexico* 28 (1963): 29–178.

15. Mittermeier, Myers, and Mittermeier, *Hotspots*, p. 89.

16. Marcos Sobral and John Renato Stehmann, "An Analysis of New Angiosperm Species Discoveries in Brazil (1990–2006)," *Taxon* 58, no. 1 (February 2009): 227–32.

17. Mittermeier, Myers, and Mittermeier, *Hotspots*, pp. 149–57.

18. For a recent short review, see Alexandre Antonelli and Isabel Sanmartin, "Why Are There So Many Plant Species in the Neotropics?" *Taxon* 60, no. 2 (April 2011): 403–14.

19. Gordon H. Orians and Antoni V. Milewski, "Ecology of Australia: The Effects of Nutrient-Poor Soils and Intense Fires," *Biological Reviews* 82, no. 3 (August 2007): 393–423.

20. Robert K. Robbins and Paul A. Opler, "Butterfly Diversity and Preliminary Comparison with Bird and Mammal Diversity," in *Biodiversity II: Understanding and Protecting Our Biological Resources*, ed. Marjorie L. Reaka-Kudla, Don E. Wilson, and Edward O. Wilson (Washington, DC: Joseph Henry Press, 1997), pp. 69–82.

21. P. Morat and P. Lowry, "Floristic Richness in the Africa-Madagascar Region: A Brief History and Prospective," *Adansonia* 19 (1997): 101–10.

22. Ronell R. Klopper et al., "Floristics of the Angiosperm Flora of Sub-Saharan Africa: An Analysis of the African Plant Checklist and Database," *Taxon* 56, no 1 (February 2007): 201–208. The numbers in this compilation seems a bit generous, though it is the best estimate we currently have for any large tropical region.

23. Andrew S. Cohen et al., "Ecological Consequences of Early Late Pleistocene Megadroughts in Tropical Africa," *Proceedings of the National Academy of Sciences of the United States of America* 104, no. 42 (October 16, 2007): 16422–27. See also: Christopher A. Scholz et al., "East African Megadroughts Between 135 and 75 Thousand Years Ago and Bearing on Early-Modern Human Origins," *Proceedings of the National Academy of Sciences of the United States of America* 104, no. 42 (October 16, 2007): 16416–21.

24. Ghillean Prance, "A Comparison of the Efficacy of Higher Taxa and Species Numbers in the Assessment of Biodiversity in the Neotropics," *Philosophical Transactions of the Royal Society* B 345, no. 1311 (July 29, 1994): 89–99.

25. Steven Goodman, personal communication (2009) based on work done by the Missouri Botanical Garden and its collaborators.

26. More detailed information can be found in a 1,700 page compendium edited by Steven M. Goodman and Johnathan P. Benstead, *The Natural History of Madagascar* (Chicago, IL: University of Chicago Press, 2003). Incidentally, a recent paper suggests that Madagascar has many more species of frogs than previously thought; see David R. Vieites et al., "Vast Underestimation of

Madagascar's Biodiversity Evidenced by an Integrative Amphibian Inventory," *Proceedings of the National Academy of Sciences of the United States of America* 106, no. 20 (May 19, 2009): 8267–72.

27. Some zoologists have speculated that the ancestors of kiwis were larger birds, like their moa cousins, and, as the bird's bodies became smaller over evolutionary time, the eggs simply didn't keep pace. Thus, kiwis have unusually large eggs for their size.

28. For an insightful and well-written review of the importance of islands in biological science, see David Quammen, *The Song of the Dodo: Island Biogeography in an Age of Extinction* (New York: Scribner, 1996).

CHAPTER 5: PATTERNS, HOTSPOTS, AND THE GEOGRAPHY OF LINEAGES

1. Anthony R. Bean, "A New System for Determining which Plant Species Are Indigenous in Australia," *Australian Systematic Botany* 20, no. 1 (2007): 1–43.

2. Arne Mooers, "The Diversity of Biodiversity," *Nature* 445, no. 7129 (February 15, 2007): 717–18. See also, Miles Spathelf and T. A. Wiate, "Will Hotspots Conserve Extra Primate and Carnivore Evolutionary History?" *Diversity and Distribution* 13, no. 6 (November 2007): 746–51.

3. Jed A. Fuhrman et al., "A Latitudinal Diversity Gradient in Planktonic Marine Bacteria," *Proceedings of the National Academy of Sciences of the United States of America* 105, no. 22 (June 3, 2008): 7774–78.

4. Attila Kalmar and David J. Currie, "A Global Model of Island Biogeography," *Global Ecology and Biogeography* 15, no. 1 (January 2006): 72–81.

5. L. A. Dyer et al. "Host Specificity of Lepidoptera in Tropical and Temperate Forests," *Nature* 448, no. 7154 (August 9, 2007): 696–99.

6. Some recent papers have questioned Rapoport's rule; the American tropics have both more narrowly and more widely distributed tree species than do temperate areas in the Western Hemisphere. See: Michael D. Weiser et al., "Latitudinal Patterns of Range Size and Species Richness of New World Woody Plants," *Global Ecology and Biogeography* 16, no. 5 (September 2007): 679–88.

7. Jason T. Weir and Dolph Schluter, "The Latitudinal Gradient in Recent Speciation and Extinction Rates in Birds and Mammals," *Science* 315, no. 5818 (March 16, 2007): 1574–76.

8. In fact things have been getting worse over the last thirty million years, since the warmth of the Eocene gave way to general cooling and greater seasonality. See: S. Bruce Archibald et al, "Seasonality, the Latitudinal Gradient of Diversity, and Eocene Insects," *Paleobiology* 36, no. 3 (June 2010): 374–98.

9. Hong Qian, Jason D. Fridley, and Michael W. Palmer, "The Latitudinal Gradient of Species-Area Relationships for Vascular Plants of North America," *American Naturalist* 170, no. 5 (November 2007): 690–701.

10. James E. Watkins Jr. et al., "Species Richness and Distribution of Ferns along an Elevational Gradient in Costa Rica," *American Journal of Botany* 93, no. 1 (January 2006): 73–83.

11. Angela Nivia Ruiz and Alfredo Cascante Marin, "Distribucion de las formas de vida en la flora costaricensis," *Brenesia* 69 (2008): 1–17.

12. Alwyn H. Gentry and C. H. Dodson, "Diversity and Biogeography of Neotropical Vascular Epiphytes," *Annals of the Missouri Botanical Garden* 74, no. 2 (1987): 205–33.

13. Jürgen Kluge and Michael Kessler, "Fern Endemism and Its Correlates: Contribution from an Elevational Transect in Costa Rica," *Diversity and Distributions* 12, no. 5 (September 2006): 535–45.

14. Jan Beck and Vun Khen Chey, "Explaining the Elevational Diversity Pattern of Geometrid Moths from Borneo: A Test of Five Hypotheses," *Journal of Biogeography* 35, no. 8 (August 2008): 1452–64.

15. Hans ter Steege et al., "Hyperdominance in the Amazonian Tree Flora," *Science* 342, no. 6156 (October 18, 2013): 325.

16. Alfred Wegener, *The Origin of Continents and Oceans*, translated from the fourth revised German edition by John Biram (New York: Dover Publications, 1966).

17. Naomi Oreskes provides a detailed analysis of why North American geologists were so reluctant to accept Wegener's theory in *The Rejection of Continental Drift: Theory and Method in American Earth Science* (New York: Oxford University Press, 1999).

18. William Glen gives a technical account of how a variety of independent research programs contributed to a final synthesis for plate tectonics in *The Road to Jaramillo: Critical Years of the Revolution in Earth Science* (Stanford, CA: Stanford University Press, 1982). A more general discussion is found in Tjeerd van Andel, *New Views on an Old Planet: A History of Global Change* (Cambridge, UK: Cambridge University Press, 1994).

19. Recent genetic evidence suggests that the ratites flew to distant regions

and became flightless only later. For a terrific example of an animal lineage still hanging on to its ancient Gondwana homeland, see Sarah L. Boyer and Gonzalo Giribet, "A New Model Gondwanan Taxon: Systematics and Biogeography of the Harvestman Family Pettalidae (Arachnida, Opiliones, Cyphophthalmi), with a Taxonomic Revision of Genera from Australia and New Zealand," *Cladistics* 23, no. 4 (August 2007): 337–61.

20. For a fine short review, see chapter 8 in Mark V. Lomolino, Brett R. Riddle, and James H. Brown, *Biogeography* (Sunderland, MA: Sinauer Assoc., 2006), pp. 227–74.

21. Ze-Long Nie et al., "Evolution of Biogeographic Disjunction between Eastern Asia and Eastern North America in *Phryma* (Phyrmaceae)," *American Journal of Botany* 93, no. 9 (September 2006): 1343–56.

22. Jun Wen, "Evolution of Eastern Asian and Eastern North American Disjunct Distributions in Flowering Plants," *Annual Review of Ecology and Systematics* 30 (1999): 421–55; Richard Milne, "Northern Hemisphere Plant Disjunctions: A Window on Tertiary Land Bridges and Climate Change?" *Annals of Botany* 98, no. 3 (September 2006): 465–72.

23. This book was published in 1999; a short scientific article outlining the reasons for designating the same twenty-five "hotspots" was published in 2000: Norman Myers et al., "Biodiversity Hotspots for Conservation Priorities," *Nature* 403, no. 6772 (February 24, 2000): 853–58.

24. Peter Kareiva and Michelle Marvier, "Conserving Biodiversity Coldspots," *American Scientist* 91, no. 4 (June 2003): 344–51.

25. V. Obando, *Biodiversidad en Costa Rica: Estao del Conociemento y Gestion* (Santo Domingo de Heredia, Costa Rica: INBio, 2002).

26. To get a better idea of the animals and plants of Costa Rica, check out these two recent guides: Carrol L. Henderson, *Field Guide to the Wildlife of Costa Rica* (Austin, TX: University of Texas Press, 2002); and Willow Zuchowski, *Tropical Plants of Costa Rica: A Guide to Native and Exotic Flora* (Ithaca, NY: Cornell University Press, 2007).

27. Edward O. Wilson, *The Diversity of Life* (Cambridge, MA: Belknap / Harvard University Press, 1992), p. 204.

28. John Terborgh estimated 15,000 plant species at the Manu Reserve, but this seems wildly exaggerated. He is quoted in Kim MacQuarrie, *Peru's Amazonian Eden: Manu National Park and Biosphere Reserve* (Barcelona, Spain: Francis O. Patthey y hijos, 1998).

29. Bruce D. Patterson, Douglas F. Stotz, and Segio Solari, "Mammals and Birds of the Manu Biosphere Reserve, Peru," *Fieldiana: Zoology* n.s., 110 (2006): 1–49.

30. Vincent S. Smith, Jessica E. Light, and Lance A. Durden, "Rodent Louse Diversity, Phylogeny, and Cospeciation in Manu Biosphere Reserve, Peru," *Biological Journal of the Linnaean Society* 95, no. 3 (November 2008): 598–610.

31. Robert K. Robbins et al., "Taxonomic Composition and Ecological Structure of the Species-Rich Butterfly Community at Pakitza, Parque Nacional del Manu, Peru," in *Manu: The Biodiversity of Southeastern Peru*, ed. Don E. Wilson and Abelardo Sanoval (Washington, DC: Smithsonian Institution, 1996), pp. 217–52.

32. Victor R. Morales and Roy W. McDiarmid, "Annotated Checklist of the Amphibians and Reptiles of Pakitza, Manu National Park Reserve Zone, with Comments on the Herpetofauna of Madre de Dios, Peru," in *Manu: The Biodiversity of Southeastern Peru*, pp. 503–22.

33. Russell A. Mittermeier, Norman Myers, and Cristina G. Mittermeier, *Hotspots: Earth's Biologically Richest and Most Endangered Terrestrial Ecoregions* (Mexico City: CEMEX, 1999), pp. 319–34.

34. Ibid., pp. 279–90.

35. Ibid., pp. 297–304.

36. Ibid., pp. 309–15.

37. Ibid., pp. 353–59.

38. Marcelo F. Simon et al., "Recent Assembly of the Cerrado, a Neotropical Plant Diversity Hotspot, by In Situ Evolution of Adaptations to Fire," *Proceedings of the National Academy of Sciences of the United States of America* 106, no. 48 (December 1, 2009): 20359–64.

39. Mittermeier, Myers, and Mittermeier, *Hotspots*, pp. 219–26.

40. Peter Goldblatt and John Manning, *Cape Plants: A Conspectus of the Cape Flora of South Africa*, vol. 9 of *Strelitzia* (St. Louis: Missouri Botanical Garden Press, 2000).

41. Holger Kreft and Walter Jetz, "Global Patterns and Determinants of Vascular Plant Diversity," *Proceedings of the National Academy of Sciences of the United States of America* 104, no. 14 (April 3, 2007): 5925–30.

42. Andrew M. Latimer, John A. Silander Jr., and Richard M. Cowling, "Neutral Ecological Theory Reveals Isolation and Rapid Speciation in a Biodiversity Hot Spot," *Science* 309, no. 5741 (September 9, 2005): 1722–25.

43. Peter Goldblatt, "An Analysis of the Flora of Southern Africa: Its Characteristics, Relationships, and Origins," *Annals Missouri Botanical Garden* 65, no. 2 (1978): 369–436.

44. Ben H. Warren and Julie A. Hawkins, "The Distribution of Species Diversity across a Flora's Component Lineages: Dating the Cape's 'Relicts,'" *Proceedings of the Royal Society B* 273, no. 1598 (September 2006): 2149–58.

45. R. M. Cowling and A. T. Lombard, "Heterogeneity, Speciation/Extinction History and Climate: Explaining Regional Plant Diversity in the Cape Floristic Region," *Diversity and Distributions* 8, no. 3 (May 2002): 163–79.

46. The notion that islands can promote the survival of species contradicts the MacArthur and Wilson equilibrium model of island biogeography, in which species numbers are controlled by a balance of immigration and local extinction. See Lawrence R. Heaney, "Is a New Paradigm Emerging for Oceanic Island Biogeography?" *Journal of Biogeography* 34, no. 5 (May 2007): 753–57.

47. See, for example, Derek Wildman et al., "Genomics, Biogeography, and the Diversification of Placental Mammals,". *Proceedings of the National Academy of Sciences of the United States of America* 104, no. 36 (September 4, 2007): 14395–400.

CHAPTER 6: SUSTAINING LOCAL BIODIVERSITY

1. D. R. Strong, "Insect Species Richness: Hispine Beetles of *Heliconia latispatha*," *Ecology* 58, no. 3 (Late Spring 1977): 573–82.

2. R. H. Whittaker, "Evolution of Species Diversity in Plant Communities," *Evolutionary Biology* 10 (1977): 22–23.

3. P. J. Grubb, "The Maintenance of Species Richness in Plant Communities: The Importance of the Regeneration Niche," *Biological Reviews* 52, no. 1 (February 1977): 119.

4. Robert John et al., "Soil Nutrients Influence Spatial Distributions of Tropical Tree Species," *Proceedings of the National Academy of Sciences of the United States of America* 104, no. 3 (January 16, 2007): 864–69.

5. J. W. Ferry Slik et al., "Environmental Correlates for Tropical Tree Diversity and Distribution Patterns in Borneo," *Diversity and Distributions* 15, no. 3 (May 2009): 523–32.

6. Kenneth P. Dial, Erick Greene, and Duncan J. Irschick, "Allometry of Behavior," *Trends in Ecology and Evolution* 23, no. 7 (July 2008): 394–401.

7. This riverside sequence was described to me by tropical forest ecologist Robin Foster. See: Robin B. Foster, Javier Arce B., and Tatzyana S. Wachter, "Dispersal and Sequential Plant Communities in Amazonian Peru Floodplain," in *Frugivores and Seed Disperal*, ed. Alejandro Estrada and Theodore H. Fleming, vol. 15 of *Tasks for Vegetation Science* (Dordrecht, Netherlands: Junk Publishers, 1986), pp. 357–70.

8. F. Stephen Dobson, "A Lifestyle View of Life-History Evolution," *Proceedings of the National Academy of Sciences of the United States of America* 104, no. 45 (November 6, 2007): 17565–66.

9. Tibor Bukovinszky et al., "Direct and Indirect Effects of Resource Quality on Food Web Structure," *Science* 319, no. 5864 (February 8, 2008): 804–807.

10. J. H. Vandermeer, J. Stout, and G. Miller, "Growth Rates of *Welfia Georgii, Socratea durissima,* and *Iriartea gigantea* Under Various Conditions in a Natural Rain Forest in Costa Rica," *Principes* 18 (1974): 148–54.

11. S. P. Hubbell and R. B. Foster, "Biology, Chance and History and the Structure of Tropical Rain Forest Tree Communities," in *Community Ecology*, ed. Jared Diamond and T. J. Case (New York: Harper and Row, 1986), pp. 314–19.

12. Peter B. Adler et al., "Climate Variability has a Stabilizing Effect on the Coexistence of Prairie Grasses," *Proceedings of the National Academy of Sciences of the United States of America* 103, no. 34 (August 22, 2006): 12793–98.

13. Christopher Wills et al., "Nonrandom Processes Maintain Diversity in Tropical Forests," *Science* 311, no. 5760 (January 27, 2006): 527–31.

14. Daniel H. Janzen, "Herbivores and the Number of Tree Species in Tropical Forests," *American Naturalist* 104, no. 940 (November/December 1970): 501–28.

15. Stephen J. Hubbell, "Seed Predation and the Coexistence of Tree Species in Tropical Forests," *Oikos* 35, no. 2 (October 1980): 214–19. Hubbell is the author of a theory that claims chance alone can explain tropical forest diversity. Called the "neutral theory," this idea counters the notion that rain forest biodiversity is primarily due to so many species having so many "different niches."

16. William J. Ripple and Robert L. Beschta, "Wolves and the Ecology of Fear: Can Predation-Risk Structure Ecosystems?" *BioScience* 54, no. 8 (August 2004): 755–66. Virginia Morell provides a more intricate update in: "Lessons from the Wild Lab," *Science* 347, no. 6228 (March 20, 2015): 1303–307. For a more general and alarming review, see: James A. Estes et al., "Trophic Downgrading of Planet Earth," *Science* 333, no. 6040 (July 15, 2011): 301–306.

17. C. K. Augspurger and C. K. Kelly, "Pathogen Mortality of Tropical Tree Seedlings: Experimental Studies on the Effects of Dispersal, Distance, Seedling Density, and Light Conditions," *Oecologia* 61, no. 2 (February 1984): 211–17.

18. Peter Thompson, *Seeds, Sex, and Civilization: How the Hidden Life of Plants Has Shaped Our World* (New York: Thames and Hudson, 2010), p. 77.

19. James M. Cook and Jean-Yves Rasplus, "Mutualists with Attitude: Coevolving Fig Wasps and Figs," *Trends in Ecology and Evolution* 18, no. 5 (May 2003): 241–48. For more about the complexity of figs, see: Sumner Silveus, Wendy Clement, and George D. Weiblen, "Cophylogeny of Figs, Pollinators, Gallers, and Parasitoids," in *Specialization, Speciation, and Radiation: The Evolutionary Biology of Herbivorous Insects*, ed. Kelley J. Tilmon (Berkeley, CA: University of California Press, 2008), pp. 225–37.

20. John Terborgh, *Diversity and the Tropical Rain Forest* (New York: Scientific American Library, 1992), p. 178. This book fails to discuss higher elevation "cloud forests."

21. Eric J. Tepe, Michael A. Vincent, and Linda E. Watson, "The Importance of Petiole Structure on Inhabitability by Ants in *Piper* sect. *Macrostachys* (Piperaceae)," *Botanical Journal of the Linnean Society* 153, no. 2 (February 2007): 181–91.

22. Ulrich Mueller and Christian Rabeling, "A Breakthrough Innovation in Animal Evolution," *Proceedings of the National Academy of Sciences of the United States of America* 105, no. 14 (April 8, 2008): 5287–88; Ted R. Schulz and Sean G Brady, "Major Evolutionary Transitions in Ant Agriculture," *Proceedings of the National Academy of Sciences of the United States of America* 105, no. 14 (April 8, 2008): 5435–40.

23. Elizabeth Pennisi, "Body's Hardworking Microbes Get Some Overdue Respect," *Science* 330, no. 6011 (December 17, 2010): 1619.

24. Luis M. Márquez et al., "A Virus in a Fungus in a Plant: Three-Way Symbiosis Required for Thermal Tolerance," *Science* 315, no. 5811 (January 26, 2007): 513–15.

25. Irwin M. Brodo, Syvia Duran Sharnoff, and Stephen Sharnoff, *Lichens of North America* (New Haven, CT: Yale University Press, 2001).

26. Robert Lucking, personal communication, Field Museum, 2007. See also Robert Lucking et al., "A First Assessment of the Ticolichen Biodiversity Inventory in Costa Rica: The Genus *Graphis*, with Notes on the Genus *Hemithecium* (Ascomycota: Ostropales: Graphidiaceae)," *Fieldiana: Botany* n.s. 46 (July 2008).

27. Yuichi Hongoh et al., "Genome of an Endosymbiont Coupling N² Fixation to Cellulolysis within Protist Cells in Termite Gut," *Science* 322, no. 5904 (November 14, 2008): 1108–109.

28. Zaal Kikvidze and Ragan M. Callawy, "Ecological Facilitation May Drive Major Evolutionary Transitions," *BioScience* 59, no. 5 (2009): 399–404. But mutualistic systems may become more vulnerable, whether in our financial systems or in nature; see: George Sugihara and Hao Ye, "Cooperative Network Dynamics," *Nature* 458, no. 7241 (April 23, 2009): 979–80.

29. This point and many other aspects of mutualism are discussed by E. G. Leigh Jr. in "The Evolution of Mutualism," *Journal of Evolutionary Biology* 23, no. 12 (December 2010): 2507–28.

30. For a broad discussion of dynamic ecological systems, see Menno Schilthuizen, *The Loom of Life: Unravelling Ecosystems* (Berlin: Springer Verlag, 2008).

31. Thom Van Dooren, *Flight Ways: Life and Loss at the Edge of Extinction* (New York: Columbia University Press, 2014), p. 37.

32. Phyllis D. Coley and Thomas A. Kursar, "On Tropical Forests and Their Pests," *Science* 343, no. 6166 (January 3, 2014): 35–36.

33. E. O. Wilson, *The Diversity of Life* (Cambridge, MA: Harvard University Press, 1992), p. 199.

34. John N. Thompson, *The Geographic Mosaic of Coevolution* (Chicago: University of Chicago Press, 2005), p. 97.

CHAPTER 7: THE EXPANSION OF BIODIVERSITY ON PLANET EARTH

1. Some elements and many different isotopes are unstable, breaking down (decaying) over time. Rate of decay is specific for each isotope's transformation, ranging from a few seconds to hundreds of millions of years. Because these changes take place within the nucleus of the atom, the rates are invariant; unaffected by pressure, temperature, or surrounding chemistry. Analyzing decayed isotopes locked within a specific rock allows geochemists to estimate the rock's age of formation. Thus, isotope decay rates have been essential in dating the origin of our planet and specific periods during its long history.

2. Matthew E. Clapham, Guy M. Narbonne, and Sames G. Gehling,

"Paleoecology of the Oldest Known Animal Communities: Ediacaran Assemblages at Mistaken Point, Newfoundland," *Paleobiology* 29, no. 4 (Fall 2003): 527–44.

3. For a detailed and well-illustrated review of the Ediacarans, see: Mikhail A. Fedonkin et al., *The Rise of Animals: Evolution and Diversification of the Kingdom Animalia* (Baltimore, MD: John Hopkins University Press, 2008).

4. Andrew H. Knoll, *Life on a Young Planet: The First Three Billion Years of Evolution on Earth* (Princeton, NJ: Princeton University Press, 2003), p. 179.

5. Martin Brasier and Jonathan Antcliffe, "Decoding the Ediacaran Enigma," *Science* 305, no. 5687 (August 20, 2004): 1115–16.

6. Lee R. Kump, "The Rise of Atmospheric Oxygen," *Nature* 451, no. 7176 (January 17, 2008): 277–78.

7. D. A. Fike et al., "Oxidation of the Ediacaran Ocean," *Nature* 444, no. 7120 (December 7, 2006): 744–47.

8. Though he devotes little attention to a rise in oxygen pressure, a good recent overview is given by Jeffrey S. Levinton, "The Cambrian Explosion: How Do We Use the Evidence?" *BioScience* 58, no. 9 (October 2008): 855–64.

9. Mark Newman, "A New Picture of Life's History on Earth," *Proceedings of the National Academy of Sciences of the United States of America* 98, no. 11 (May 22, 2001): 5955–56.

10. Oliver Morton, *Eating the Sun: How Plants Power the Planet* (New York: HarperCollins, 2008), p. 233.

11. Jane Gray and William Shear, "Early Life on Land," *American Scientist* 80, no. 5 (September/October 1992): 444–56.

12. Diane Ackerman, *Cultivating Delight: A Natural History of My Garden* (New York: HarperCollins, 2001), p. 221.

13. Brigitte Meyer-Berthaud and Anne-Laure Decombeux, "A Tree Without Leaves," *Nature* 446, no. 7138 (April 19, 2007): 861–62.

14. Dimitrios Floudas et al., "The Paleozoic Origin of Enzymatic Lignin Decomposition Reconstructed from 31 Fungal Genomes," *Science* 336, no. 6089 (June 29, 2012): 1715–19.

15. Two fine technical reviews are: Burkhard Becker and Birger Marin, "Streptophyte Algae and the Origin of Embryophytes," *Annals of Botany* 103, no. 7 (May 2009): 999–1004; and Yin-Long Qiu, "Phylogeny and Evolution of Charophytic Algae and Land Plants," *Journal of Systematics and Evolution* 46, no. 3 (2008): 287–306.

16. Some estimates run the total number of flowering plants up to about

400,000 species. But this is not the number actually described at this time. Also, I prefer the lower figure of 260,000 because of the problem of synonomy, where many species carry multiple names or have been split into unnecessary segregate species.

17. Some Gymnosperms are pollinated by insects, but even these usually build seeds in advance of fertilization.

18. C. Kevin Boyce et al., "Angiosperm Leaf Vein Evolution was Physiologically and Environmentally Transformative," *Proceedings of the Royal Society B* 276, no. 1663 (May 22, 2009): 1771–76.

19. Harald Schneider et al, "Ferns Diversified in the Shadow of Angiosperms," *Nature* 428, no. 6982 (April 1, 2004): 553–56.

20. Corrie S. Moreau et al., "Phylogeny of the Ants: Diversification in the Age of Angiosperms," *Science* 312, no. 5770 (April 7, 2006): 101–104.

21. Kim Roelants et al., "Global Patterns of Diversification in the History of Modern Amphibians," *Proceedings of the National Academy of Sciences of the United States of America* 104, no. 3 (January 7, 2007): 887–92.

22. Better than this, I claimed that without the flowering plants human beings and their fancy societies simply wouldn't be here; see, *Flowers: How They Changed the World* (Amherst, NY: Prometheus Books, 2006).

23. Why the dinosaurs, especially some of the herbivorous ones, became so large is an intriguing question. Even the largest mammals of the last fifty million years did not get close to the size of either the large herbivorous dinosaurs or to carnivorous *Tyrannosaurus* and its allies. See: P. Martin Sander and Marcus Clauss, "Sauropod Gigantism," *Science* 322, no. 5899 (October 10, 2008): 200–201.

24. Important references are: L. W. Alvarez et al., "Extraterrestrial Cause for the Cretaceous-Tertiary Extinction," *Science* 208, no. 4448 (June 6, 1980): 1095–1108; Walter Alvarez, *T. Rex and the Crater of Doom* (Princeton, NJ: Princeton University Press, 1997); and William Glen, ed., *The Mass Extinction Debate: How Science Works in a Crisis* (Stanford, CA: Stanford University Press, 1994).

25. For an accessible account of the end-Permian extinction, see Douglas H. Erwin, *Extinction: How Life on Earth Nearly Ended 250 Million Years Ago* (Princeton, NJ: Princeton University Press, 2006). An important recent technical analysis is: Ezat Heydari, Nasser Arzani, and Jamshid Hassanzadeh, "Mantle Plume: The Invisible Serial Killer—Application to the Permian-Triassic Boundary Mass Extinction," *Paleogeography, Paleoclimatology, Paleoecology* 264, no. 1–2 (July 7, 2008): 147–62.

26. After a long period of warmth, the world began to cool around 34 mya (at the Eocene-Oligocene boundary), and ice sheets began to grow in Antarctica. See Gabriel Bowen, "When the World Turned Cold," *Nature* 445, no. 7128 (February 8, 2007): 607–608. From this point on, temperatures worldwide decline, culminating in the ice ages of the last two million years. A recent hypothesis claims that the rise of the Himalayas played a major role in cooling the world. See: E. Irving, "Why Earth Became So Hot 50 Million Years Ago and Why It Then Cooled," *Proceedings of the National Academy of Sciences of the United States of America* 105, no. 42 (October 21, 2008): 16061–62.

27. Niles Eldredge, *Life Pulse: Episodes from the Story of the Fossil Record* (New York: Facts on File, 1987).

28. Jennifer C. McElwain and Surangi W. Punyasena, "Mass Extinction Events and the Plant Fossil Record," *Trends in Ecology and Evolution* 22, no. 10 (October 2007): 548–56.

29. Many mammals became distinctly larger and more diverse after the end-Cretaceous extinction, see: Felisa A. Smith et al., "The Evolution of Maximum Body Size of Terrestrial Mammals," *Science* 330, no. 6008 (November 26, 2010): 1216–19.

30. Norman F. Hughes, *Paleobiology of Angiosperm Origins* (Cambridge, UK: Cambridge University Press, 1976), p. 36.

31. Karl J. Niklas, Bruce H. Tiffney, and Andrew H. Knoll, "Patterns of Vascular Land Plant Diversification," *Nature* 303, no. 5918 (June 16, 1983): 614–16.

32. Storrs L. Olson, "Why so Many Kinds of Passerine Birds?" *BioScience* 51, no. 4 (April 2001): 268–69. See also: Peter M. Bennett and Ian P. F. Owens, *Evolutionary Ecology of Birds* (Oxford, UK: Oxford University Press, 2002).

33. Michael J. Benton and Brent C. Emerson, "How Did Life Become So Diverse? The Dynamics of Diversification According to the Fossil Record and Molecular Phylogenetics," *Paleontology* 50, no. 1 (January 2007): 23–40.

34. J. J. Sepkoski, "A Factor Analytic Description of the Phanerozoic Marine Fossil Record," *Paleobiology* 7, no. 1 (Winter 1981): 36–53; Andrew M. Bush and Richard J. Bambach, "Did Alpha Diversity Increase during the Phanerozoic? Lifting the Veils of Taphonomic, Latitudinal, and Environmental Biases," *Journal of Geology* 112, no. 6 (November 2004): 625–42.

35. John Alroy et al., "Phanerozoic Trends in the Global Diversity of Marine Invertebrates," *Science* 321, no. 5885 (July 4, 2008): 97–100.

36. Peter J. Wagner, Mathew A. Rosnik, and Scott Lidgard, "Abundance

Distributions Imply Elevated Complexity of Post-Palaeozoic Marine Eco-systems," *Science* 314, no. 5803 (November 24, 2006): 1289–92. Also see the commentary by Wolfgang Kiessling in the same issue: 1254–55.

37. For a negative view regarding increased marine biodiversity over geological time, see: Patrick D. Wall et al., "Revisiting Raup, Exploring the Influence of Outcrop Area on Diversity in Light of Modern Sample-Standardization Techniques," *Paleobiology* 35, no. 1 (November 2009): 146–47.

38. James S. Crampton et al., "The Ark Was Full! Constant to Declining Cenozoic Shallow Marine Biodiversity on an Isolated Midlatitude Continent," *Paleobiology* 32, no. 4 (2006): 509–32.

39. An authoritative and up-to-date overview on land diversity is Michael Benton's "The Origin of Modern Biodiversity on Land," *Philosophical Transactions of the Royal Society B* 365, no. 1558 (November 27, 2010): 3667–79.

CHAPTER 8: A WORLD OF EVER-INCREASING COMPLEXITY

1. Eric D. Beinhocker, *The Origin of Wealth: Evolution, Complexity and the Radical Remaking of Economics* (Boston, MA: Harvard Business School Press, 2006), p. 18. I have replaced *natural selection* for the author's *evolution* in this quotation; this is a common error. An important book, but the author has little to say about the energy needed to drive further economic growth or its environmental effects.

2. Kevin Kelly, *What Technology Wants* (New York: Viking, 2010), p. 274. A challenging text, but nothing about sustainability.

3. Steven Strogatz, *Sync: The Emerging Science of Spontaneous Order* (New York: Hyperion Books, 2003), p. 286.

4. Daniel W. McShea and Robert H. Brandon discuss their idea of continuing diversification over time in their book *Biology's First Law* (Chicago: University of Chicago Press, 2010).

5. Mitch Leslie provides a concise review: "On the Origin of Photosynthesis," *Science* 323, no. 5919 (March 6, 2009): 1286–87. For a well-written survey of photosynthesis, see Oliver Morton, *Eating the Sun: How Plants Power the Planet* (New York: HarperCollins, 2008).

6. This is paraphrased from a lecture by physicist Martin Kamen, as reported on page 125 of Oliver Morton's book *Eating the Sun*.

7. For a series of recent technical reviews, see Euan Nisbet et al, eds., "Photosynthetic and Atmospheric Evolution," *Philosophical Transactions of the Royal Society B* 363, 1504 (August 27, 2008): 2623–801.

8. Neil Shubin, *Your Inner Fish: A Journey into the 3.5 Billion Year History of the Human Body* (New York: Pantheon Books, 2008), p. 46.

9. You may prefer to see this grand epic as one created and guided by a creative Deity, and that is fine. The problem for science is that a belief in God's "intelligent design" cannot be tested. This is the fundamental difference between religious belief and scientific practice. The former is based on a tradition of revealed truth; the latter has been constructed from data we find in nature or is revealed by experiment. Religious debates are framed within accepted scripture. Scientific debates are supposed to be based on the latest data derived directly from an analysis of nature. Of course, both traditions are enlivened by arguments regarding the interpretation of scripture or the interpretation of data. For an accessible review of the scientific evidence for evolution, see Jerry Coyne's *Why Evolution Is True* (New York: Viking, 2009).

10. Sean B. Carroll, *Endless Forms Most Beautiful: The New Science of Evo-Devo and the Making of the Animal Kingdom* (New York: Norton, 2005).

11. Elizabeth Pennisi, "Working the (Gene Count) Numbers: Finally a Firm Answer?" *Science* 316, no. 5828 (May 25, 2007): 1113.

12. Nessa Carey, *Junk DNA: A Journey through the Dark Matter of the Genome* (New York: Columbia University Press, 2015).

13. For those interested in a mathematical look at the complexity of life, Ricard Solé and Brian Goodwin provide significant insights in their *Signs of Life: How Complexity Pervades Biology* (New York: Basic Books, 2000).

14. Wallace Arthur, *Creatures of Accident: The Rise of the Animal Kingdom* (New York: Hill and Wang, 2006), pp. 43, 25.

15. Keiko Sakakibara et al., "KNOX2 Genes Regulate the Haploid-to-Diploid Morphological Transition in Land Plants," *Science* 339, no. 6123 (March 1, 2013): 1067–70. See also: William E. Friedman, "One Genome, Two Ontogenies," *Science* 339, no. 6123 (March 1, 2013): 1045–46.

16. The rise of Angiosperms has been a major triumph in the history of mutualism, see: E. G. Leigh Jr., "The Evolution of Mutualism," *Journal of Evolutionary Biology* 23, no. 12 (December 2010): 2507–28.

17. The KTR concept and the importance of flowering plants are well summarized by Michael Benton in "The Origins of Modern Biodiversity on

Land," *Philosophical Transactions of the Royal Society B* 365, no. 1558 (November 27, 2010): 3667–80. See also: Michael Novacek, *Terra: Our 100-Million-Year-Old Ecosystem—and the Threats that Now Put It at Risk* (New York: Farrar, Straus and Giroux, 2007). This book includes considerable earth science. For a simpler overview, see my *Flowers: How They Changed the World* (Amherst, NY: Prometheus Books, 2006).

18. Steven Jay Gould, "Play It Again, Life," *Natural History* 95, no. 2 (February 1986): 18–26.

19. Simon Conway Morris, *Life's Solution: Inevitable Humans in a Lonely Universe* (London: Cambridge University Press, 2002).

20. Sakakibara et al., "KNOX2 Genes." See also: Friedman, "One Genome, Two Ontogenies."

21. Scott Turner argues that homeostasis is a central governing factor in the history of life; see his *The Tinkerer's Accomplice: How Design Emerges from Life Itself* (Cambridge, MA: Harvard University Press, 2007).

22. T. S. Kemp, "The Concept of Correlated Progression as the Basis of a Model for the Evolutionary Origin of Major Taxa," *Proceedings of the Royal Society B* 274, no. 1618 (July 7, 2007): 1667–73.

23. For a technical review, see: Jennifer A. Clack, *Gaining Ground: The Origin and Evolution of Tetrapods* (Bloomington, IN: Indiana University Press, 2002).

24. Darwinism suffered many other criticisms over the last century. For a short but insightful overview by a philosopher, see: R. G. Winther, "Systemic Darwinism," *Proceedings of the National Academy of Sciences of the United States of America* 105, no. 33 (August 19, 2008): 11833–838.

25. Leigh Van Valen, "A New Evolutionary Law," *Evolutionary Theory* 1 (1973): 1–30. But note that the Red Queen Hypothesis has also been used to explain the maintenance of sex; see: G. Bell, *The Masterpiece of Nature: The Evolution and Genetics of Sexuality* (Berkeley, CA: University of California Press, 1982).

26. David M. Raup, *Extinction: Bad Genes or Bad Luck?* (New York: Norton, 1991).

27. For a recent discussion regarding animal camouflage, see: Martin Stevens and Sami Merilaita, "Animal Camouflage: Current Issues and New Perspectives," *Philosophical Transactions of the Royal Society B* 364, no. 1516 (February 27, 2009): 423–27, and related articles.

28. John T. Bonner, *The Evolution of Complexity by Means of Natural Selection* (Princeton, NJ: Princeton University Press, 1988).

29. For a taste of astrobiology and the argument that we might find life within the icy crust of Jupiter's moon Europa, see: Marc Kaufman, *First Contact: Scientific Breakthroughs in the Hunt for Life beyond Earth* (New York: Simon and Schuster, 2011).

30. Geerat J. Vermeij, *Evolution and Escalation: An Ecological History of Life* (Princeton, NJ: Princeton University Press, 1987).

31. John Warren Huntley and Michal Kowalewski, "Strong Coupling of Predation Intensity and Diversity in the Phanerozoic Fossil Record," *Proceedings of the National Academy of Sciences of the United States of America* 104, no. 38 (September 18, 2007): 15006–10.

32. For a short overview, see: John J. Flynn, "Splendid Isolation," *Natural History* 118, no. 5 (June 2009): 26–32. A more detailed review is by S. David Webb, "The Great American Biotic Interchange: Patterns and Process," *Annals of the Missouri Botanical Garden* 93, no. 2 (August 2006): 245–57.

33. Paul S. Martin, *Twilight of the Mammoths: Ice Age Extinctions and the Rewilding of America* (Berkeley, CA: University California Press, 2005).

34. Bert Hölldobler and E. O. Wilson, *The Superorganism: The Beauty, Elegance and Strangeness of Insect Societies* (New York: W. W. Norton, 2008).

35. Ted R. Schultz and Sean G. Brady, "Major Evolutionary Transitions in Ant Agriculture," *Proceedings of the National Academy of Sciences of the United States of America* 105, no. 14 (April 8, 2008): 5435–40.

36. The concept of "group selection" has been wrongly rejected for more than three decades. See the arguments by David Sloan Wilson in "Rethinking the Theoretical Foundation of Sociobiology," *Quarterly Review of Biology* 82, no. 4 (December 2007): 327–46.

37. Edward O. Wilson, "One Giant Leap: How Insects Achieved Altruism and Colonial Life," *BioScience* 58, no. 1 (January 2008): 17–25. See also, William O. H. Hughes et al., "Ancestral Monogamy Shows Kin Selection Is Key to the Evolution of Sociality," *Science* 320, no. 5880 (May 30, 2008): 1213–16.

38. Zhe-Xi Luo, "Transformation and Diversification in Early Mammal Evolution," *Nature* 450, no. 7172 (December 13, 2007): 1011–19.

39. Robert W. Meredith et al., "Impacts of the Cretaceous Terrestrial Revolution and End-Cretaceous Extinction on Mammal Diversification," *Science* 334, no. 6055 (October 28, 2011): 521–24.

40. Maureen A. O'Leary et al., "The Placental Mammal Ancestor and Post-K-Pg Radiation of Placentals," *Science* 339, no. 6120 (February 8, 2013): 662–69.

41. Juliane Kaminski et al., "Word Learning in a Domestic Gog: Evidence for 'Fast Mapping,'" *Science* 304, no. 5677 (June 11, 2004): 1682–83.

42. Unfortunately, the large majority of fossils are of marine life, but they also give evidence for increasing diversity over time; see: Steven M. Stanley, "An Analysis of the History of Marine Animal Diversity," *Paleobiology* 33, no. S4 (Fall 2007): 1–55.

43. Though biological science has focused on negative and competitive interactions in the history of life, a strong case can be made for positive effects as well. See: Zaal Kikvidze and Ragan M. Callaway, "Ecological Facilitation May Drive Major Evolutionary Transitions," *BioScience* 59, no. 5 (May 2009): 399–404.

44. W. Brian Arthur, "On the Evolution of Complexity," in *Complexity: Metaphors, Models, and Reality*, ed. G. Cowan, D. Pines, and D. Meltzer, vol. 19 of *Santa Fe Institute Studies in the Sciences of Complexity* (Reading, MA: Addison Wesley, 1994), pp. 65–81.

45. Gerald Schönknecht et al., "Gene Transfer from Bacteria and Archae Facilitated Evolution of an Extremophilic Eukaryote," *Science* 339, no. 6124 (March 8, 2013): 1207–10.

46. W. Brian Arthur, *The Nature of Technology* (New York: Free Press, 2009), p. 10. This book is an incisive review of human technology and its elaboration over time.

47. Dirk K. Morr, "Lifting the Fog of Complexity," *Science* 343, no. 6169 (January 24, 2014): 382–83.

48. Karen M. Kaphelm et al., "Genomic Signatures of Evolutionary Transitions from Solitary to Group Living," *Science* 348, no. 6239 (June 5, 2015): 1139–42. A study of bee species informs our own evolution.

49. Lizzie Wade, "Birth of the Moralizing Gods," *Science* 349, no. 6251 (August 28, 2015): 919–22. Large societies required strong moralistic religions.

CHAPTER 9: BIOLOGICAL COMPLEXITY TRIUMPHANT: THE HUMAN MIND

1. Diane Ackerman, *Cultivating Delight: A Natural History of My Garden* (New York: HarperCollins, 2001), p. 236.

2. Peter Ward, *The Medea Hypothesis: Is Life on Earth Ultimately Self-Destructive?* (Princeton, NJ: Princeton University Press, 2009), p. 126.

3. For a detailed review regarding how our biosphere has changed since flowering plants proliferated, see: Michael Novacek, *Terra: Our 100-Million-Year-Old Ecosystem—and the Threats That Now Put It at Risk* (New York: Farrar, Straus and Giroux, 2007).

4. Nick Lane, *Power, Sex, Suicide: Mitochondria and the Meaning of Life* (Oxford, UK: Oxford University Press, 2005), p.108.

5. Gould even wrote that progress "is a delusion based on social prejudice and psychological hope." (Stephen J. Gould, *Full House: The Spread of Excellence from Plato to Darwin* [New York: Harmony Books, 1996], p.20.) For a short critical review of this book see the Dawkins reference below.

6. Daniel W. McShea, "Metazoan Complexity and Evolution: Is There a trend?" *Evolution* 50, no. 2 (April 1996): 489. McShea provides a clear discussion of his concerns and lists a wide array of references.

7. See: Brendon M. H. Larson, "The Social Resonance of Competitive and Progressive Evolutionary Metaphors," *BioScience* 56, no. 12 (December 2006): 997–1004.

8. Seth Finnegan, Jonathan L. Payne, and Steve C. Wang, "The Red Queen Revisited: Reevaluating the Age Selectivity of Phanerozoic Marine Genus Extinctions," *Paleobiology* 34, no. 3 (2008): 318–41; Steve C. Wang and Andrew M. Bush, "Adjusting Global Extinction Rates to Account for Taxonomic Susceptibility," *Paleobiology* 34, no. 4 (2008): 434–55.

9. Wallace Arthur calls this his "lawn with molehills perspective" of life's history. The lawn represents increased species diversity without increasing complexity. The molehills are his occasional advances in complexity. See: Wallace Arthur, *Creatures of Accident: The Rise of the Animal Kingdom* (New York: Hill and Wang, 2006), p. 5.

10. Michael Ruse, *Monad to Man: The Concept of Progress in Evolutionary Biology* (Cambridge, MA: Harvard University Press, 1996), p. 146.

11. Aaron Clauset and Douglas H. Erwin, "The Evolution and Distribution of Species Body Size," *Science* 321, no. 5887 (July 18, 2008): 399–401.

12. Jonathan Payne et al., "Two-Phase Increase in the Maximum Size of Life over 3.5 Billion Years Reflects Biological Innovation and Environmental Opportunity," *Proceedings of the National Academy of Sciences of the United States of America* 106, no. 1 (January 6, 2009): 24–27.

13. Richard Dawkins, "Human Chauvinism and Evolutionary Progress," chapter 5.4 in *A Devil's Chaplain: Reflections on Hope, Lies, Science, and Love* (New

York: Mariner Books, 2004), p. 214; first published as "Human Chauvinism," *Evolution* 51, no. 3 (June 1997): 1015–20.

14. Harry Jerison, *Evolution of the Brain and Intelligence* (New York: Academic Press, 1973).

15. John Allman, *Evolving Brains*, vol. 68 in Scientific American Library (New York: Freeman, 1999).

16. See: Marcia Ponce de Leon et al., "Neanderthal Brain Size at Birth Provides Insights into the Evolution of Human Life History," *Proceedings of the National Academy of Science of the United States of America* 105, no. 37 (September 16, 2008): 13764–68.

17. A major new discovery is reported by David Lordkipanidze et al., "A Complete Skull from Dmanisi, Georgia, and the Evolutionary Biology of Early *Homo*," *Science* 342, no. 6156 (October 18, 2013): 326–31.

18. I'm sure that the "Out-of-Africa" scenario is nowhere near as simple as usually characterized. See, for example: Philipp Gunz et al, "Early Human Diversity Suggests Subdivided Population Structure and a Complex Out-of-Africa Scenario," *Proceedings of the National Academy of Sciences of the United States of America* 106, no. 15 (April 14, 2009): 6094–98.

19. G. A. Lyras et al., "The Origin of *Homo floresiensis* and Its Relation to Evolutionary Processes under Isolation," *Anthropological Science* 117, no. 1 (2009): 33–43. For a review, see: Leslie C. Aiello, "Five Years of *Homo floresiensis*," *American Journal of Physical Anthropology* 142, no. 2 (June 2010): 167–79.

20. Christopher Heesy, "Seeing in Stereo: The Ecology and Evolution of Primate Binocular Vision and Stereopsis," *Evolutionary Anthropology* 18, no. 1 (January/February 2009): 21–35.

21. R. W. Sussman, "Primate Origins and the Evolution of Angiosperms," *American Journal of Primatology* 23, no. 4 (1991): 209–23. Also: Robert W. Sussman, "How Primates Invented the Rainforest and Vice Versa," in *Creatures of the Dark: The Nocturnal Prosimians*, ed. L. Alterman et al. (New York, Plenum Press, 1995), pp. 1–9; Dario Maestripieri, *Machavellian Intelligence: How Rhesus Macques and Humans Have Conquered the World* (Chicago: University of Chicago Press, 2007), p. 159.

22. Robert Martin, *How We Do It: The Evolution and Future of Human Reproduction* (New York: Basic Books, 2013), p.124. An excellent evolutionary overview.

23. Donald Johanson and B. Edgar, *From Lucy to Language* (New York: Simon and Schuster, 1996). See also: Tim D. White et al., "*Ardipithecus ramidus*

and the Paleobiology of Early Hominids," *Science* 326, no. 5949 (October 2, 2009): 64. Part of a special issue.

24. Lucy had a foot very much like the one we walk on; see: Carol V. Ward, William H. Kimbel, and Donald C. Johanson, "Complete Fourth Metatarsal and Arches in the Foot of *Australopithecus afarensis*," *Science* 331, no. 6018 (February 11, 2011): 750–53.

25. "The human foot is our most distinctive adaptation," declare Robin H. Crompton and Todd C. Pataky in a short review: "Stepping Out," *Science* 323, no. 5918 (February 27, 2009): 1174–75.

26. Alan Walker, "The Strength of Great Apes and the Speed of Humans," *Current Anthropology* 50, no. 2 (April 2009): 229–34.

27. A few relevant references are: M. C. Corballis, "The Gestural Origin of Language," *American Scientist* 87, no. 2 (March/April 1999): 138–45; Marc Hauser and T. Bever, "A Biolinguistic Agenda," *Science* 322, no. 5904 (November 14, 2008): 1057–59; Michael Tomasello, *Origins of Human Communication* (Cambridge, MA: MIT Press, 2008).

28. Kevin Kelly, *What Technology Wants* (New York, Viking. 2010), p. 26.

29. How brains achieve cognition and perception is being examined using new mathematical and neural models. See: Misha Rabinovich, Ramon Huerta, and Giles Laurent, "Transient Dynamics for Neural Processing," *Science* 321, no. 5885 (July 4, 2008): 48–50.

30. For a broader perspective, see: Robert D. Martin, "The Evolution of Human Reproduction: A Primatological Perspective," *American Journal of Physical Anthropology* 134, no. S45 (2007): 59–84.

31. Ann Gibbons, "The Birth of Childhood," *Science* 322, no. 5904 (November 14, 2008): 1040–43.

32. Jean-Jacques Hublin, "The Prehistory of Compassion," *Proceedings of the National Academy of Sciences of the United States of America* 106, no. 16 (April 21, 2009): 6429–30.

33. Sarah Blaffer Hrdy, *Mothers and Others: The Evolutionary Origins of Mutual Understanding* (Cambridge, MA: Harvard University Press, 2009). Both a mother and astute scientist, Hrdy has written several insightful books on the evolution and sociology of mothering.

34. David P. Barash, *Natural Selections: Selfish Altruists, Honest Liars and Other Realities of Evolution* (New York: Bellevue Literary Press, 2008), p. 62.

35. J. M. Burkart, S. B. Hrdy, and C. P. Van Schak, "Cooperative Breeding and

Human Cognitive Evolution," *Evolutionary Anthropology* 18, no. 5 (September/October 2009): 175–86. For a short review, see: Elizabeth Pennisi, "On the Origin of Cooperation," *Science* 325, no. 5945 (September 4, 2009): 1196–99.

36. Richard Potts, *Humanity's Descent: The Consequences of Ecological Instability* (New York: Wm. Morrow, 1996).

37. David Berreby, *Us and Them: The Science of Identity* (Chicago: University of Chicago Press, 2008), p.327. Richly documented, this is an important text. Politically correct on occasion, Berreby never relates "race" to geography and claims, incorrectly, that DNA studies do not support the concept.

38. For a discussion regarding the origins of human conflict, see: Laurent Lehmann and Marcus W. Feldman, "War and the Evolution of Belligerence and Bravery," *Proceedings of the Royal Society B* 275, no. 1653 (December 22, 2008): 2877–85.

39. The reluctance of humanists to address our warlike and nasty behavior is surely a reflection of intellectual fashion. One of the few recent studies on human nastiness is Kathleen Taylor's *Cruelty: Human Evil and the Human Brain* (New York: Oxford University Press, 2009).

40. Y. Fernández-Jalvo et al., "Evidence of Early Cannibalism," *Science* 271, no. 5247 (January 19, 1996): 277–78. These finds are similar to earlier work in southern France; see: P. Villa et al., "Cannibalism in the Neolithic," *Science* 233, no. 4762 (July 25,1986): 431–37. For a short review, see: Tim D. White, "Once Were Cannibals," *Scientific American* 284, no. 1 (January 2001): 58–65.

41. E. O. Wilson, quoted by Amanda Ruggeri in *U.S. News and World Report* December 15, 2008, p. 16.

42. Richard D. Alexander, "The Evolution of Social Behavior," *Annual Review of Ecology and Systematics* 5 (1974): 325–83; Richard D. Alexander, *The Biology of Moral Systems* (Hawthorne, NY: Aldine De Gruyter, 1987).

43. Charles Darwin, *The Descent of Man and Selection in Relation to Sexual Selection*, second revised ed. (New York: D. Appleton and Co. 1896), p.132.

44. Samuel Bowles, "Conflict: Altruism's Midwife," *Nature* 456, no. 7220 (November 20, 2008): 326. Charles Darwin expressed very similar views, as cited in the previous reference.

45. Samuel Bowles, "Did Warfare among Ancestral Hunter-Gatherers Affect the Evolution of Human Social Behaviors?" *Science* 324, no. 5932 (June 5, 2009): 1293–98. See also: David Berreby, *Us and Them: the Science of Identity* (Chicago: University of Chicago Press, 2008).

46. Actually, human brain expansion was already well underway before the *sapiens-neanderthal* split—during the time of *Homo erectus*, 1.8 mya. See: Scott W. Simpson et al, "A Female *Homo erectus* Pelvis from Gona, Ethiopia," *Science* 322, no. 5904 (November 14, 2008): 1089–92.

47. Francis Thackeray, in a lecture at Field Museum, April 2009. See also: Juli G. Pausas and Jon E. Keeley, "A Burning Story: The Role of Fire in the History of Life," *BioScience* 59, no. 7 (July/August 2009): 593–601.

48. For more in this regard, see: Richard Wrangham, *Catching Fire: How Cooking Made Us Human* (New York: Basic Books, 2009).

49. For a fine review of human energy use, see: Alfred W. Crosby, *Children of the Sun: A History of Humanity's Unappeasable Appetite for Energy* (New York: W.W. Norton, 2006).

50. Hillard S. Kaplan, Paul L. Hooper, and Michael Gurven, "The Evolutionary and Ecological Roots of Human Social Organization," *Philosophical Transactions of the Royal Society B* 364, no. 1533 (November 12, 2009): 3289–99.

51. For a fine discussion of gender differences and how they can be ameliorated, see: Lise Eliot, *Pink Brain, Blue Brain: How Small Differences Grow Into Troublesome Gaps—And What We Can Do About It* (Boston, Houghton Mifflin Harcourt, 2009).

52. Joshua Greene, *Moral Tribes: Emotion, Reason and the Gap between Us and Them* (New York: Penguin Press, 2013). Another significant text is: Edward O. Wilson, *The Social Conquest of Earth* (New York: W. W. Norton, 2012).

53. Erik Trinkhaus, "European Early Modern Humans and the Fate of the Neanderthals," *Proceedings of the National Academy of Sciences of the United States of America* 104, no. 18 (May 1, 2007): 7367–72.

54. Richard E. Green et al., "A Draft Sequence of the Neandertal Genome," *Science* 328, no. 5979 (May 7, 2010): 710–22.

55. Thomas Wynn, "Hafted Spears and the Archaeology of Mind," *Proceedings of the National Academy of Sciences of the United States of America* 106, no. 24 (June 16, 2009): 9544–45; Lyn Wadly, Tamaryn Hodgskiss, and Michael Grant, "Implications for Complex Cognition from the Hafting of Tools with Compound Adhesives in the Middle Stone Age, South Africa," *Proceedings of the National Academy of Sciences of the United States of America* 106, no. 24 (June 16, 2009):9590–94. The earliest dates for stone spear points are close to 500,000 years ago: Jayne Wilkins et al., "Evidence for Early Hafted Hunting Technology," *Science* 338, no. 6109 (November 16, 2012): 942–46.

56. Kenny Smith et al, "Introduction: Cultural Transmission and the Evolution of Human Behavior," *Philosophical Transactions of the Royal Society B* 363, no. 1509 (November 12, 2008): 3471. An introduction to a special issue of ten papers, this article is, in part, an effort to counter the claims of "evolutionary psychology" with cultural dynamics. Seems to me that both approaches are valid and cannot really be disentangled.

57. Actually something like Lamarckian evolution (epigenetics), where experience changes heredity, is found in bacterial grade organisms. Epigenetics is also a current focus of research in higher, eukaryotic organisms, like ourselves.

58. Ludwig Huber et al, "The Evolution of Imitation: What Do the Capacities of Non-Human Animal Tell Us about the Mechanisms of Imitation?" *Philosophical Transactions of the Royal Society B* 364, no. 1528 (August 27, 2009): 2293–98.

59. Ignore those who declare that climate change or an asteroid impact wiped out much of the megafauna of North America 11,000 years ago. Instead, read: Paul S. Martin, *Twilight of the Mammoths: Ice Age Extinctions and the Rewilding of America* (Berkeley, CA: University of California Press, 2005), but ignore "rewilding."

CHAPTER 10: EVER MORE COMPLEXITY: HUMAN CULTURAL ADVANCE

1. James Lovelock, *The Revenge of Gaia: Earth's Climate Crisis and the Fate of Humanity* (New York: Basic Books, 2007), p. 15.

2. Paul R. Ehrlich and Anne H. Ehrlich, *The Dominant Animal: Human Evolution and the Environment* (Washington, DC: Island Press, 2008), p. 240.

3. Jeffrey D. Sachs, *Common Wealth: Economics for a Crowded Planet* (New York: Penguin Books, 2008), p. 58.

4. Anthropologists often refer to this dramatic innovation as the Neolithic Revolution. For a good review, see: Bruce D. Smith, *The Emergence of Agriculture* (New York: Scientific American Library, 1995). For more recent technical views, see: Tim Denham, Jose Iriarte, and Luc Vrydaghs, eds. *Rethinking Agriculture: Archaeological and Ethnoarcaeological Perspectives* (Walnut Creek, CA: Left Coast Press, 2007).

5. Some think that it was the ending of the ice age itself that helped initiate

agriculture; see, for example: Oliver Morton, *Eating the Sun: How Plants Power the Planet* (New York: Harper, Collins Publisher, 2008), p. 301.

6. Geerat J. Vermeij, "Escalation and Its Role in Jurassic Biotic History," *Paleogeography, Paleoclimatology, Paleoecology* 263, no. 1–2 (June 13, 2008): 3.

7. Jonathan A Foley et al., "Our Share of the Planetary Pie," *Proceedings of the National Academy of Sciences of the United States of America* 104, no. 31 (July 31, 2007): 12585–86.

8. For a fine review of trading in human history, see: William J. Bernstein, *A Splendid Exchange: How Trade Shaped the World* (New York: Atlantic Monthly Press, 2008).

9. Ara Norenzayan and Azim F. Shariff, "The Origin and Evolution of Religious Prosociality," *Science* 322, no. 5898 (October 3, 2008): 58–62.

10. Jared Diamond, *Guns, Germs and Steel: The Fates of Human Societies* (New York: W. W. Norton, 1997). A fine overview, but there is no suggestion that iron and steel technology may have been a uniquely Middle Eastern innovation.

11. Keith Parsons, *It Started with Copernicus: Vital Questions about Science* (Amherst, NY: Prometheus Books, 2014), p. 197.

12. Harold J. Berman, *Law and Revolution: The Formation of the Western Legal Tradition* (Cambridge, MA: Harvard University Press, 1983). By developing a new legal system, separate from religious dictates, Europe set itself free in a way that other societies had not.

13. Patrick Alitt, *A Climate of Crisis. America in the Age of Environmentalism* (New York: Penguin Press, 2014), p. 10.

14. "The price of firewood in Britain rose 700% between 1500 and 1630." (Alfred W. Crosby, *Children of the Sun: A History of Humanity's Unappeasable Appetite for Energy* [New York: W. W. Norton, 2006], p. 69.)

15. For a fine review of energy utilization in both the natural and human world, see: Vaclav Smil, *Energy in Nature and Society* (Cambridge, MA: MIT Press, 2008).

CHAPTER 11: A FOUR-BILLION-YEAR EPIC

1. Neil Shubin gives a fine account of the evidence found within our own bodies that attests to our long evolutionary ancestry, see: *Your Inner Fish: A Journey into the 3.5 Billion-Year History of the Human Body* (New York: Pantheon Books, 2008).

2. Franklin M. Harold, *In Search of Cell History* (Chicago: University of

Chicago Press, 2014), p.75. A technical but masterful review of how cells may have arisen.

3. Christian de Duve discusses these ideas with clarity and depth in *Vital Dust: Life as a Cosmic Imperative* (New York: Basic Books, 1995).

4. For a detailed and well-written survey of how we have come to understand photosynthesis, as well as its effect on the biosphere over time, see: Oliver Morton, *Eating the Sun: How Plants Power the Planet* (New York: HarperCollins, 2008).

5. We reviewed the importance of mitochondria in chapter 2, and how endosymbiosis created a new platform for bio-complexity.

6. Yuval Noah Harari, *Sapiens: A Brief History of Humankind* (New York: HarperCollins, 2015), pp. 3–39. An insightful analysis!

7. Martin A. Nowak, *Super Cooperators: Altruism, Evolution and Why We Need Each Other to Succeed* (New York: Free Press, 2011), p. xiv.

8. Paul Wapner, *Living Through the End of Nature: The Future of American Environmentalism* (Cambridge, MA: MIT Press, 2010), p. xiv. For more views on our culture's representations of "nature," see: William Cronon, ed., *Uncommon Ground: Toward Reinventing Nature* (New York: W. W. Norton, 1995).

9. Gaia Vince, *Adventures in the Anthropocene: A Journey to the Heart of a Planet We Made* (Minneapolis, MN: Milkweed Editions, 2014), p. 67. Travelling around the world, the author has put together a very informative book. However, statements like "Humans have the power to heat the planet further or cool it right down" undermine her optimistic narrative.

CHAPTER 12: TRILLIONS OF TRANSISTORS:
AN UNCERTAIN FUTURE

1. Edward O. Wilson, *The Social Conquest of Earth* (New York: W. W. Norton, 2012), p. 13.

2. Paul K. Conkin, *The State of the Earth: Environmental Challenges on the Road to 2100* (Lexington, KY: University Press of Kentucky, 2007), p. 281.

3. W. Brian Arthur, *The Nature of Technology: What It Is and How It Evolves* (New York: Free Press, 2009), p. 10. This book is a brilliant analysis of technology and its elaboration over time. Unfortunately, the author has nothing to say regarding how technology might threaten our biosphere.

4. Kevin Kelly, *What Technology Wants* (New York: Viking, 2010), p. 187.

A magisterial overview you ought to read, though energy consumption and resource diminution simply aren't there.

5. Ibid., p. 143. The idea that our genes might be adapting to our modern age ignores the fact that really smart and successful people have relatively low birthing rates.

6. Many of my recent numbers regarding population growth are from *Wikipedia* and from a special section regarding population in *Science* 333, no. 6042 (July 29, 2011): 529–86.

7. Robert B. Laughlin, *Powering the Future: How We Will (Eventually) Solve the Energy Crisis and Fuel the Civilization of Tomorrow* (New York: Basic Books, 2011). For a somewhat different perspective, see: Scott L. Montgomery, *The Powers That Be: Global Energy in the Twenty-First Century and Beyond* (Chicago: University of Chicago Press, 2010).

8. Lionel Tiger, *Optimism: The Biology of Hope* (New York: Simon and Schuster, 1979), pp. 15, 21.

9. Tali Sharot, *The Optimism Bias: A Tour of the Irrationally Positive Brain* (New York: Pantheon Books, 2011), p. 22.

10. Julian Simon, ed., *The State of Humanity* (Cambridge, MA: Blackwell Publishers, 1995).

11. Thomas L. Friedman, *The World Is Flat: A Brief History of the Twenty-First Century*, first rev. ed. (New York: Farrar, Strauss, Giroux, 2006). I cite this book because it reflects journalism's inattention to the population explosion.

12. Eric Roston, *The Carbon Age: How Life's Core Element Has Become Civilization's Greatest Threat* (New York: Walker and Co., 2008), p. 188.

13. Ben J. Wattenberg, "The Population Explosion Is Over," *New York Times Magazine*, 23 November 1997: 60–63.

14. Alexander Skutch, *Life Ascending* (Austin, TX: University of Texas Press, 1985), p. 197.

15. Peter Gill, *Famine and Foreigners: Ethiopia Since Live Aid* (New York: Oxford University Press, 2010), p. 123.

16. The two thick but demography-deficient books are Ross Garnaut, *The Garnaut Climate Change Review* (Cambridge, UK: Cambridge University Press, 2008); and Nicholas Stern, *The Economics of Climate Change* (Cambridge, UK: Cambridge University Press, 2007).

17. Thomas R. Malthus, *An Essay on the Principal of Population, as it Affects the Future Improvement of Society* (London: J. Johnson, 1798).

18. Sabrina Tavernise, "Survey of Pakistan's Young Predicts 'Disaster' if Their Needs Aren't Addressed," *New York Times, International,* November 22, 2009, p. 15.

19. Ross Gelbspan, *Boiling Point: How Politicians, Big Oil and Coal, Journalists and Activists Are Fueling the Climate Crisis and What We Can Do to Avert the Disaster* (New York: Basic Books, 2004), p.70.

20. Mark Van Putten, "How to Save a Planet: A Users Guide," *BioScience* 58, no. 9 (October 2008): 874–79.

21. David E. Bloom, "7 Billion and Counting," *Science* 333, no. 6042 (July 29, 2011): 562–69. This issue features a special section on population.

22. V. Ramanathan and Y. Feng, "On Avoiding Dangerous Anthropogenic Interference with the Climate System: Formidable Challenges Ahead," *Proceedings of the National Academy of Sciences of the United States of America* 105, no. 38 (September 23, 2008): 14245–50.

23. Peter Ward, *The Medea Hypothesis: Is Life on Earth Ultimately Self-Destructive?* (New Jersey: Princeton University Press, 2008). Ward's thesis, I fear, is correct, though I disagree with a number of his opinions.

24. Jared Diamond, *Collapse: How Societies Choose to Fail or Succeed* (New York: Viking, 2005). Vaclav Smil dismisses Diamond's thesis, even citing a paper claiming that Easter Island's demise was due to an invasion of rats, followed by the influx of Western diseases. I see Smil's attitude as another example of our eagerness to deny negative human agency. See reference 29, below.

25. Simon A. Levin, "Self-Organization and the Emergence of Complexity in Ecological Systems," *BioScience* 55, no. 12 (December 2005): 1075–79.

26. Deborah MacKenzie, "What Price More Food?" *New Scientist,* 2660 (June 14, 2008): 28–33.

27. Peter Dauvergne, *The Shadows of Consumption: Consequences for the Global Environment* (Cambridge, MA: MIT Press, 2008), p. 5.

28. Jeffrey D. Sachs, *Common Wealth: Economics for a Crowded Planet* (New York: Penguin Press, 2008), p.139. This is an excellent overview, which remains optimistic despite awesome obstacles.

29. Vaclav Smil, *Global Catastrophes and Trends: The Next 50 Years* (Cambridge, MA: MIT Press, 2008). Smil has published a number of highly informative books on global issues.

30. Brahma Chellaney, *Water, Peace, and War: Confronting the Global Water Crisis* (Lanham, MD: Rowan and Littlefield, 2013).

31. For a thorough account of how we got to the place we are and why population growth is so great a problem, see: Paul R. Ehrlich and Anne H. Ehrlich, *The Dominant Animal: Human Evolution and the Environment* (Washington, DC: Island Press, 2008).

32. Thomas L. Friedman, *Hot, Flat, and Crowded: Why We Need a Green Revolution and How It Can Renew America* (New York: Farrar, Straus and Giroux, 2008), p. 186. Unfortunately, more and more growth is utterly unsustainable, which explains why I am so pessimistic about our technological civilization's future: core human values must change, but that seems unlikely.

33. Bill McKibben, *Eaarth: Making a Life on a Tough New Planet* (New York: Henry Holt, 2011), pp. 27, 125.

34. Richard Heinberg, *The Party's Over: Oil, War and the Fate of Industrial Societies* (Gabriola Island, BC: New Society, 2005).

35. Laurence C. Smith, *The World in 2050: Four Forces Shaping Civilization's Northern Future* (New York: Dutton, 2010), p. 261. This book surveys many ways in which we impact our planet.

36. Elizabeth Kolbert, *Field Notes from a Catastrophe: Man, Nature, and Climate Change* (New York: Bloomsbury, 2006), p. 187. See also her more recent volume: *The Sixth Extinction: An Unnatural History* (New York: Henry Holt, 2014).

37. Thomas Merton, *Watch for the Light: Readings for Advent and Christmas* (New York: Maryknoll, 2004), p. 276.

38. Daniel Alpert, *The Age of Oversupply: Overcoming the Greatest Challenge to the Global Economy* (New York: Penguin Publishers, 2013), p. 200.

39. The "length of life of industrial societies" was the last factor in Frank Drake's famous equation, estimating how many radio-broadcasting civilizations might exist in our Milky Way Galaxy. See the final chapter in my book, *Perfect Planet, Clever Species: How Unique Are We?* (Amherst, NY: Prometheus Books, 2003).

40. Philip Smith and Nicolas Howe, *Climate Change as Social Drama: Global Warming in the Public Sphere* (New York: Cambridge University Press, 2015), p. 3. This book examines many aspects of human societies, which will, I believe, make any concerted solutions unlikely.

41. Roger V. Short, "Population Growth in Retrospect and Prospect," *Philosophical Transactions of the Royal Society B* 164, no. 1532 (October 27, 2009): 2971.

42. Alan Weisman, *Countdown: Our Last Best Hope for a Future on Earth?* (New York: Little, Brown, 2013), p. 306. Having visited many societies around the globe, Weisman reports on how our species is affecting the planet: a good read!

INDEX